ENERGY
SELF-SUFFICIENCY

THE AEI
NATIONAL ENERGY PROJECT

The American Enterprise Institute's
National Energy Project was established in early 1974
to examine the broad array of issues
affecting U.S. energy demands and supplies.
The project will commission research into all important
ramifications of the energy problem—economic
and political, domestic and international, private
and public—and will present the results
in studies such as this one.
In addition it will sponsor symposia, debates, conferences,
and workshops, some of which will be televised.

The project is chaired by Melvin R. Laird,
former congressman, secretary of defense,
and domestic counsellor to the President,
and now senior counsellor of *Reader's Digest*.
An advisory council, representing a wide range of
energy-related viewpoints, has been appointed.
The project director is Professor Edward J. Mitchell
of the University of Michigan.

Views expressed are those of the authors
and do not necessarily reflect the views of
either the advisory council and others associated with
the project or of the advisory panels,
staff, officers, and trustees of AEI.

ENERGY SELF-SUFFICIENCY
An economic evaluation

M.I.T. Energy Laboratory Policy Study Group

American Enterprise Institute for Public Policy Research
Washington, D. C.

This study was prepared by the M.I.T. Energy Laboratory Policy Study Group: M. A. ADELMAN,* professor of economics; ROBERT E. HALL, associate professor of economics; KENT F. HANSEN, professor of nuclear engineering; J. HERBERT HOLLOMON, professor of engineering; HENRY D. JACOBY,* professor of management; PAUL L. JOSKOW,* assistant professor of economics; PAUL W. MACAVOY,* professor of management; HERMAN P. MEISSNER,* professor of chemical engineering, emeritus; DAVID C. WHITE,* Ford professor of engineering and director of the Energy Laboratory; and MARTIN B. ZIMMERMAN,* research associate in management. (*Asterisks indicate principal authors of the report.*)

The Policy Study Group was assisted by: Donald B. Anthony, Martin Baughman, Joseph Bell, Edward Erickson, Richard Gordon, Robert P. Greene, James Gruhl, Jerry A. Hausman, Edward Hudson, Dale Jorgenson, William A. Litle, Richard Mancke, James W. Meyer, Charles M. Mohr, James Sloss, and Robert Spann.

ISBN 0-8447-3144-7

National Energy Study 3, November 1974

Library of Congress Catalog Card No. 74-23886

An earlier version of this monograph appeared in the May 1974 *Technology Review*, edited at the Massachusetts Institute of Technology.
© 1974 by the Alumni Association of M.I.T.

Printed in the United States of America

CONTENTS

FOREWORD TO
THE AEI EDITION

This monograph, the third in the National Energy Studies series, is a new departure for us. An earlier version of this study was published in the May 1974 issue of *Technology Review*, which is edited at the Massachusetts Institute of Technology. Subsequent to its publication, conversations between the authors and me led to the conclusion that an updated version might be in order in light of rapidly changing events and advances in the M.I.T. Energy Laboratory's economic modeling efforts. Further work by the authors resulted in significant changes in some of the quantitative projections made in the earlier studies. It is hoped that publication of this revised version will enable the energy policy maker and the informed citizen to keep abreast of the latest findings in this vital research field.

While, as I noted above, many of the numerical forecasts have been revised, the basic conclusion of the paper remains the same: Achieving U.S. energy self-sufficiency by the 1980s would almost certainly mean much higher prices for American consumers than a policy that relied on some imported oil. In other words, the premium paid for national energy insurance would exceed the cost of the damages the insurance is supposed to protect us against.

Edward J. Mitchell
Project Director

FOREWORD

The Report and Its Sources

This report had its genesis at least four years ago, when a number of members of the M.I.T. faculty became concerned with this nation's ever-growing use of energy. Given limits on domestic supplies, it was evident even then that changes in the extent and nature of energy availability, however they came, would soon deeply and broadly affect the American economy.

Analysis of the energy situation was expanded in M.I.T. classrooms and laboratories, with the result that a great deal of new work was begun in many fields of technology, management, and the social sciences. To crystallize and provide a focus for the interdisciplinary interests which were clearly emerging, the Institute established an Energy Laboratory in 1972; its director is David C. White, Ford professor of engineering.

Thus by the fall of 1973, when it became clear that the forecasts of energy-induced dislocations might be fulfilled far more quickly than most of us had expected, the Institute had in being an informed and effective working group with an understanding of energy issues ranging from economics, marketing, and policy problems to technological options and environmental issues. The M.I.T. administration believed this to be a uniquely informed, independent resource, and we accordingly proposed early this year that the Energy Laboratory's Policy Study Group turn its attention to the concept of "Project Independence"—the effort to develop and exploit U.S. energy resources so intensively that by 1980 the nation would no longer need to depend on imported fuels. The implications of that effort—the question of whether the goal could be achieved and, if so, the price its achievement might entail for the nation—became the focus of a major M.I.T. study, of which the following is the full, final report. The group and its consultants from other universities have been greatly encouraged and helped by the interaction of many government agencies whose staffs shared data and reviewed or helped refine the interim results.

The report which follows does not propose to show how the nation can solve its energy problems. Its central theme is to study the responsiveness of today's complex energy system to those changes in the supply of fuels and in the demand for energy which are in fact possible by 1980 through present and foreseeable technology, and the effect on both these of changes in the prices of fuel and of energy. The best of forecasters is humbled by questions in this realm, and yet the issues now before the country demand that our discussion of policies and alternatives be based on the most competent and disinterested understanding of just these interrelationships. The purpose of this report, and of M.I.T. in sponsoring it, is to help achieve that goal.

Behind every energy bottleneck and in every future decision stand serious societal issues—nuclear power plant safety, environmental protection, and many others. Such issues, though both appropriate and important to the debate which is now in progress throughout the nation, are beyond the scope of this report.

Albert G. Hill
Vice President for Research, M.I.T.

1

INTRODUCTION

Overview of the Study

The United States is dedicated to a policy of independence from foreign sources of energy. Under the name "Project Independence," goals have been set which call for making the nation invulnerable to oil embargoes or price increases in the world petroleum market. As originally presented, this goal was specified in terms of complete self-sufficiency in energy by the end of the decade. In recent months, however, the feasibility of this particular objective has been called into question, and as part of a larger-scale study of a "blueprint" for Project Independence, the Federal Energy Administration (FEA) is attempting to formulate an operational definition of "independence" and develop a set of measures to achieve it.

This study seeks to contribute to discussion of this issue by evaluating the economic implications of an attempt to achieve self-sufficiency in this critical commodity. Taking 1980 as a target year, forecasts are made of U.S. energy demand and of the domestic supplies of various forms of energy, and the results are used to predict the prices at which domestic supply and demand will be in equilibrium. The results indicate that prices of $11.00 to $13.00 per barrel (oil equivalent) will be necessary to bring forth enough additional supplies of fossil fuels to satisfy demands in domestic energy markets by that time.[1] The picture is little changed if the target date is moved to 1985.

---

[1] All price figures are stated in terms of *constant 1973 prices*. To convert to nominal prices of later years, the reader must apply his own inflation factor. For example, to convert to price levels as of late 1974, a factor of 10 to 15 percent must be added.

This means that, even if concerted efforts were made to remove the bottlenecks that exist in these markets (such as the federal price regulation of natural gas or various environmental limitations), there would have to be another round of price increases for consumers roughly as great as that experienced in 1973-1974. In short, self-sufficiency, as a form of "insurance" against disruption or price increase, will be purchased at a very high cost. The curtailment of imports acts to replace a temporary embargo, or a threat of an embargo, with a permanent embargo that increases prices beyond present levels.

There are other ways to approach the problem of U.S. dependence on foreign sources. One way involves the use of flexible tariffs or quotas to hold the oil price at a high level that would reduce imports. This method is examined in detail, and it is concluded that the current oil price is high enough to extract present domestic oil and gas reserves with great efficiency. A still higher price would have only a marginal effect on exploration and production over the next few years within the United States. Current prices also provide ample incentives for coal production. However, it might require a doubling of price to provide enough incentive to bring about large-scale commercial development of synthetic fuels in the near future, and their development is not sufficiently promising of large supplies to justify such high prices *for all energy*. It is concluded, therefore, that the import price now prevailing is high enough, and little is to be gained by raising it more. The use of tariffs or quotas to increase prices should not be adopted as an instrument of policy at this time.

For many of the same reasons, there is no need at present to provide a firm floor under present prices. Most industry experts seem to believe that most of the exploration and development activity that would take place at $9.00 to $10.00 per barrel will also occur at prices as low as $7.00 per barrel, so there is little need to set a price floor under current conditions. Should estimates of supply costs change, this judgment would have to be reconsidered.

Special price policies should be developed for the synthetics industry, though. The best way to encourage significant supplies of "syngas," "syncrude," methanol, or shale oil in the *late* nineteen-eighties would be to identify these as a special class of new energy sources and provide specific price guarantees for such output. The federal government ought to offer to purchase oil, gas, or methanol from the first few synthetic commercial plants at a price agreed upon by negotiation or resulting from competitive bids. Naturally, we should avoid an open-ended commitment to subsidize this infant industry, for the costs of these sources, in relation to other supplies, are not yet known.

Security can be provided against import disruption by the introduction of radically new import policies. One important element would be an import storage program, by which a stockpile of crude oil beyond normal inventories would be maintained as a hedge against significant and lengthy embargoes. The maintenance of a stockpile to guard against a one-year cutoff of 2 million barrels per day of imports would cost about $990 million per year. If the government required that it be provided by the oil industry, the cost of oil delivered to consumers would rise by no more than twenty-five cents per barrel, or two-thirds of a cent per gallon. Other elements of a foreign purchase program are not worked out in this study, but the general thrust of the scheme would be to purchase oil from abroad when the sources are forecast to be free from "bloc" policies to restrict supplies. While independence can be partially achieved a number of different ways, the least-cost solution will most likely involve a combination of several specific import and subsidy schemes, rather than a complete cessation of imports.

The Critical Uncertainties

In the wake of the recent Mideast war and the disruption of international petroleum markets, a commitment has been made to free the United States from dependence on insecure and expensive foreign sources of energy. The program to achieve this objective has been termed Project Independence. As originally defined in the midst of the oil embargo, it implied a goal of domestic self-sufficiency by 1980. Subsequent analysis has led to a softening of this definition to allow for a more leisurely target deadline or to introduce the notion that there may be some level of continuing imports that is sufficiently small that the economy is not threatened by a cut-off. Whatever the definition, however, Project Independence raises issues that cut across the full range of government and corporate policies in the energy sector of the economy. There are implications for controls of domestic prices, controls of imports, for incentives to domestic production, and for the development of new technologies. Most important, there are implications for the prices which American consumers will pay for energy or for products using energy—which in turn embrace the full range of goods and services comprising the U.S. gross national product (GNP).

The central concern is whether Project Independence can result in domestic supply and demand equilibrium at socially tolerable prices.

Examination of the issues requires a comparison of the likely market-clearing prices for energy products under two conditions: complete self-sufficiency and the absence of self-sufficiency, which implies some degree of net import demand. Such forecasts are extremely difficult to make at the present time, because of critical and in some cases irreducible uncertainty in five sets of factors.

One: Responsiveness of Domestic Supply and Demand to Price Changes. There is now available a wide range of estimates of expected demand and supply for energy in the early nineteen-eighties. Some studies show the market for most energy products clearing by 1980 with no net imports (or nearly none) on the basis of an (equivalent) oil price of $6.00 to $7.00 per barrel. Other studies indicate that the market-clearing price will rise as high as $15.00 per barrel before self-sufficiency is attained.

The desirable mix of policy measures to achieve Project Independence depends upon whether the price is $6.00 or $15.00 per barrel, and the economic costs and benefits of Project Independence are greatly different over this range of outcomes. If the economy were to adjust smoothly to exclusively domestic supply at $6.00 per barrel, then the desired government policy would be to remove unnecessary bottlenecks to the adjustment process and otherwise do nothing. But if a price of $15.00 is required to clear markets with exclusively domestic supplies, then the policy of doing nothing would have undesirable income distribution implications and perhaps unacceptable effects on the rate of growth and employment of resources in the domestic economy. Therefore, it is extremely important to reduce the uncertainties in predictions of the 1980 equilibrium price, insofar as it is possible to do so.

The approach here is to combine studies of supply for specific energy resources with overall demand forecasts to estimate "total supply and total demand" for energy in the domestic economy. The studies used imply different levels of responsiveness of supply and demand to price, and the resulting estimates define a range of prices likely under energy self-sufficiency. Chapters 3 through 5 are devoted to specific fuels, and the overall supply and demand estimates for 1980 are summarized in Chapter 2.

Two: World Oil Price. The price of oil imports to the United States is now well over $9.00 per barrel. Evaluation of Project Independence requires an estimate of this price in the future, because the alternative to self-sufficiency is to continue to purchase supplies from interna-

4

tional oil markets. If the world oil price were greatly reduced, the cost of Project Independence—in terms of our inability to purchase cheap supplies abroad—would be greatly increased.

There is little basis upon which to predict any specific oil price over the next ten years. As discussed in Chapter 6, the critical uncertainty concerns the ability of the Persian Gulf states to restrict production so as to increase the price for their sales and the sales of other countries not taking part in output restriction. These political matters cannot be forecast with any degree of accuracy. As a result, the price could drift up or down over a period of years within a range of $4.00 to $12.00 per barrel, and it becomes impossible to evaluate Project Independence with precision. All that can be done here is to concentrate our analysis on the range of possible quantities and prices of oil and gas from abroad.

Three: The Costs of Synthetic Fuels. Sooner or later, liquefaction or gasification of coal will take place in the United States, and there will be substantial development of oil-shale resources. The issue is not whether such resources will be forthcoming, but when there will be capacity sufficient to provide substantial volumes of oil-equivalent supplies. The timing is not easy to estimate. Any assessment depends on the trade-off between the pace of development and its cost in terms of resources and environmental damage. This trade-off is not well-defined at present; while several liquefaction or gasification technologies are under active development, they involve untested technology. Where experience with particular processes has become available, operation has taken place at a scale from 10 to 30 percent of the size of contemplated commercial units.

Estimates of the price incentives necessary to bring these new processes on-line, and at capacity levels supporting production of millions of barrels per day, can be made only within very wide ranges. Attempts are made in Chapter 5 to summarize data currently available on the processes and establish the ranges of likely costs for each. While they do not promise significant supplies by the early nineteen-eighties at prices comparable to those for fossil fuels, there may be some chance that these processes can make at least modest additions to domestic energy resources in this period.

Four: Expansion Capacity of the Construction Industry. In the process of scaling up the domestic fuel industry, tremendous pressures will be placed on particular segments of the construction industry. Prices charged for construction might rise steeply as a result, and critical bottlenecks could develop in completing the schedule for achieving independence in the nineteen-eighties. These problems in construction may be exacerbated by problems in transportation of fuels: for example, there may not be sufficient capacity to provide unit-train service for coal from the western regions of the country. The expansion capabilities of the construction industry, the transportation industries, and providers of industrial inputs such as the water system of the United States are not well understood at the present time. Further careful consideration of the "scaling" effect on industries serving the energy sector of the economy is required.

Five: The Nature of Security. Foreign sources of energy are highly diverse in the security problems they present. The level of American vulnerability is a complex function of the total import volume, the fraction of imports from any one country, and the specific sources of imports. As a result, there is critical uncertainty about the dimensions and the likelihood of the disruptive events that Project Independence is trying to guard against. This makes it difficult to evaluate Project Independence, since the nature of the security gain is not well specified.

Uncertainties in these five areas make it extremely difficult to analyze *any* policy favoring self-sufficiency in energy. But there can be no doubt that Project Independence will require much higher levels of domestic prices than are likely to occur without independence. This will be shown in the next chapter in summary form and in the following chapters in some detail.

2
ENERGY SUPPLY AND DEMAND IN 1980

When trade takes place with foreign suppliers, the price of energy can never rise above the price at which supply would equal demand if only domestic supplies were available, since importation from abroad must add some quantity to supply, and thus reduce the price. In fact, the difference between the "world trade" price and an "exclusively domestic" price is an indication of the costs—that is, the additional amounts paid per unit of consumption—that would result from invoking a policy of self-sufficiency. As a first step in finding such costs, we forecast these levels of prices.

Tables 1 and 2 present a general picture of the domestic energy market in 1980, with all quantities converted into equivalent barrels of oil.[1] Supplies of different fuels and overall energy demand are estimated at prices of $7.00, $9.00, and $11.00 per barrel (in constant 1973 dollars), and the results define points on approximate supply and demand curves for total energy. By looking at the intersection of these curves—the point where domestic supply and demand appear to be in balance—we can predict the prices implied by a goal of self-sufficiency at the end of the decade.

Table 1 contains supply estimates based on detailed econometric studies of oil and natural gas. These analyses rely on extrapolation of the recent behavior of energy markets, and as such the forecasts for 1980 are likely to be an optimistic extension of patterns in the nineteen-sixties and early nineteen-seventies. The table also shows two estimates of demand, one econometric and one judgmental. The econometric estimate (Hudson-Jorgenson) is based upon a large-scale econometric model of energy demand in the United States, and reflects

---

[1] A fuel is made "oil equivalent" by finding the number of barrels of oil which has the same heating value as a given quantity of that fuel.

Table 1

ENERGY EQUILIBRIUM IN 1980:
SUPPLY FORECASTS BASED ON ECONOMETRIC MODELS

Fuel	Source of Estimate	Millions of Barrels Per Day Equivalent, at Prices Per Barrel		
		$7.00	$9.00	$11.00
Crude oil and natural gas liquids (including Alaskan)	M.I.T. model	10.6	10.7	10.9
		(2.0)	(2.0)	(2.0)
Natural gas	M.I.T. model	14.7	14.5	14.4
Coal	M.I.T. analysis	6.1	8.0	8.0
Uranium and hydroelectric	Equipment survey	5.2	5.2	5.2
New technology	M.I.T. analysis	0.0	0.0	0.1
Total supply		36.6	38.4	38.6
Forecasts of total demand	Hudson-Jorgenson	44.2	42.4	40.6
	Judgmental	45.6	45.6	45.6
Net imports	Hudson-Jorgenson demand less total supply	7.6	4.0	2.0

a relatively strong change in demand when prices change. The judgmental estimate does not take explicit account of the response of energy demand to price changes.

Table 2 shows the same two demand estimates, but it provides a contrasting prediction of supply, based on judgmental forecasts for oil and natural gas instead of analytical model results. These judgmental forecasts were made by individuals or organizations versed in the energy industry; their "inputs" included not only formal or quantitative modeling of recent experience but qualitative analysis of the future as well. As a comparison of the tables shows, the econometric and judgmental forecasts of oil and gas imply similar levels of supply throughout the price range. The two demand estimates are the same at approximately $5.50 per barrel (which is below the price range shown in the tables), but the econometric estimate is below the judgmental at prices above this level.

Table 2

ENERGY EQUILIBRIUM IN 1980: JUDGMENTAL SUPPLY FORECASTS

Fuel	Source of Estimate	Millions of Barrels Per Day Equivalent, at Prices Per Barrel		
		$7.00	$9.00	$11.00
Crude oil and natural gas liquids (including Alaskan)	NPC (Case I)	13.6 (2.0)	13.6 (2.0)	13.6 (2.0)
Natural gas	NPC (Case II)	11.5	11.5	11.5
Coal	M.I.T. analysis	6.1	8.0	8.0
Uranium and hydro-electric	Equipment survey	5.2	5.2	5.2
New technology	M.I.T. analysis	0.0	0.0	0.1
Total supply		36.4	38.3	38.4
Forecast of total demand	Hudson-Jorgenson	44.2	42.4	40.6
	Judgmental	45.6	45.6	45.6
Net imports	Hudson-Jorgenson demand less total supply	7.8	4.1	2.2

The tables indicate a wide range of possible values for the price required to clear domestic markets using only United States energy sources. At one extreme, the optimistic forecast results from the econometric estimates of total supply and total demand in Table 1, which show market-clearing in the neighborhood of $11.00 to $13.00 per barrel. This is an optimistic prediction; for it to be borne out, both the econometric estimates of supply (which are probably expansive for natural gas) and the econometric model of demand (which shows a strong price response) must be assumed to hold. If the judgmental demand estimates of Table 2 prove to be more accurate, then the prohibition of imports implies a price per barrel greatly beyond the range included in the tables. Extending the judgmental forecasts into the range of $11.00 to $20.00 per barrel, our own assessment is that the high end of clearing prices would be at least as high as $14.00 per barrel.

9

A calculation by Martin Baughman, using the M.I.T. model of interfuel substitution, shows a market clearing price of $12.50 per barrel under the Hudson-Jorgenson demand forecast. Using the judgmental demand prediction, the clearing price is over $14.00 per barrel.

All these forecasts are necessarily imprecise. The judgmental estimates normally imply conservative assumptions about price response, or they ignore price altogether—a clear weakness of the judgmental approach. On the other hand, the econometric models are limited in their capacity to consider the effects of future resource depletion or constraints on supplies of equipment and personnel. Hence the judgmental and econometric methods should be used to complement each other, so in this case the truth likely falls somewhere between the low clearing price yielded by econometric methods and the high price implicit in the judgmental analysis.

Until more experience is gained at these high price levels, the uncertainty over this range of prices will remain. On balance, however, the results point to the conclusion that the price of energy would be from $11.00 to $13.00 per barrel if supplies were limited to those within the United States.

The Supply Forecasts

The economic and judgmental forecasts of supply are described in detail in Chapters 3 through 6. But here it may be appropriate to characterize them in general terms, so as to point out the extent of imprecision in the numbers shown in the tables.

The Bases of the Forecasts. The econometric analyses in Table 1 show four important sources of domestic energy: crude oil, natural gas, coal, and nuclear and hydroelectric power. Crude oil sources are forecast to produce less from onshore domestic wells in 1980 than at the present time because the incentives of increased price are not sufficient to compensate for depletion of inground reserves. This reduction is compensated by increases in offshore, Alaskan, and natural gas liquids production. The supplies of natural gas are characterized by growth over the next few years. It is assumed that field prices will not rise very much due to the regulatory actions of the Federal Power Commission (FPC), but the growth in output in response to even modest increase in price more than compensates for depletion effects.

As prices rise from $7.00 to $11.00 in the table, the estimated natural gas supply, which is based on separate models of onshore and offshore development, actually declines slightly. This is because the *substitution* of oil for gas drilling at much higher oil prices cancels out some of the expansive effects of the modest FPC-allowed price increase which is assumed for natural gas. Oil supply is augmented at the higher prices. But this estimate is based on an analysis that does not take separate account of offshore development; potential offshore expansion is treated as an extrapolation of the overall oil sector experience of recent years, and as such, the estimate very likely understates the potential for offshore supplies. An improved model of offshore oil might add as much as one million barrels per day to expected supply at the higher prices, and this correction would erase the apparent paradox (the total of oil and gas supplies does not rise as prices increase) shown in Table 1.

The supplies of coal are predicted with the assumption of an entirely new industry: strip mining in southeastern Montana. Additional supplies are available at higher prices, but at prices of $9.00 per barrel oil equivalent or above, the estimates shown in Tables 1 and 2 are limited by the country's capacity to consume coal.

There are two critical estimates which must be made in forecasting energy supplies: supply responsiveness to price increases and depletion effects. These estimates are quite imprecise, because of poor data in all energy industries—particularly in coal, where there is no industry at all in Montana at the present time.

It is more difficult to assess the judgmental forecasts reported in Table 2. Here, one usually cannot know what factors determined the estimate, nor can one consider the variation around the forecast. Thus these forecasts could be more subject to error than an economic forecast, but there is no way of knowing this. The approach here has been to use the forecasts that have been made in most detail, by individuals or organizations who have worked for some time on either the procedures for forecasting or the forecasts themselves.

The Sources of the Forecasts. The econometric forecast for oil supply is based upon a crude-oil supply model in existence for more than a decade. It was first constructed by Franklin Fisher at M.I.T., then reconstructed and updated in more recent years by Edward Erickson and Robert Spann at North Carolina State University. The last version of this model—used for the forecasts in Table 1—is a combined oil and natural gas econometric model developed at M.I.T. that has been used extensively for forecasting the results of policy change in the regulation of natural gas. The model, described below in Chapter 3, combines oil and gas forecasting in a detailed regional formulation.

11

The supplies of coal have heretofore been neglected by fore-casters; our procedure has been to use the expertise and judgment of two economic analysts working on coal at the present time—Martin Zimmerman of M.I.T. and Richard Gordon of Pennsylvania State University. They produce a combination economic-judgmental fore-cast that is used in Tables 1 and 2. Similarly, supplies from new technology are estimated by a group of engineering analysts versed in these technologies and working under the direction of Herman P. Meissner at M.I.T. Their forecasts are judgmental, but heavily bol-stered by analytical work on the performance of new technology in the recent past. Finally, supplies of uranium and hydroelectric energy are fixed by present plant capacities and by the construction of new plants in the next few years. Therefore, forecasts of such supply are limited to what these plants can produce.

In Table 2, we substitute the National Petroleum Council (NPC) judgmental forecasts for the econometric forecasts. The NPC study is noted for exceptional detail and for having reconciled a wide variety of views of those in the oil industry concerning future supplies. It is not an analytical or even a formal forecast; there is some doubt whether the NPC study even relates supplies of crude oil, gas, or coal to alternative prices. But it is used here because of the authority of those who participated in the NPC exercise—the wide range of interests and expertise was extremely impressive—and because these individuals dealt directly with *future* depletion, while the econometric models only extrapolate past depletion.

The Hudson-Jorgenson Model of Demand

Edward Hudson and Dale Jorgenson of Data Resources, Inc., attempt to forecast demands and supplies for nine broad industrial sectors over the period 1973-2000. They can then obtain a projection of total energy demand and supply over that period. Three basic models are used to provide the forecasts. First, a "long-term" macroeconomic model is used to predict levels of final GNP demand and also to predict the prices of the factors of production, capital and labor. Then, taking these final demands and prices as given, two further models—a production model and a consumer-behavior model—are used to calcu-late the inter-industry flow of products and the prices for these products.

To calculate final energy demands, Hudson and Jorgenson must forecast total demand for products and the energy used per unit of product. This computation involves a very complicated model, since

12

both total demand for products and the demand for inputs used in creating these products are determined simultaneously as a function of equilibrium prices, the prices at which markets clear. Hudson and Jorgenson simplify this problem by separating the consumption and production sides of the model through assuming an input-output structure of production. However, as an advance over previous work, they estimate the input-output coefficients as a function of prices. Thus the procedure for determining energy demand begins with final demand for products by four sectors: private consumption, private investment, government expenditures, and net exports. Given final demand, the modellers multiply by energy input per unit of final demand, which is calculated from the input-output coefficients. Then, summing across all final demand, an estimate of total energy demand is determined. As prices change, the input-output coefficients will change, and thus energy use will respond to changes in price.[2]

In Tables 1 and 2, we use the forecast of total energy consumption from the Hudson-Jorgenson model. The model shows strong price sensitivity to changes in demand, having been based on data from a period in which reduced prices were accompanied by increased consumption. These data are used for forecasting a period in which price *increases* are expected to be followed by *reduced* demands. It is therefore assumed that the demand processes observed in the past, and formalized by Hudson and Jorgenson, are reversible, although there may be some doubt about the completeness of the reversibility.

The Judgmental Demand Forecasts

Many attempts at forecasting demand have been based on projections of recent trends in energy consumption and on the forecasters' knowledge of individual industries. While such forecasts do not enable us to estimate demand responses to changes in prices, they may still be useful as "boundary" projections for the relatively near future. Such a forecast is used to predict total energy demand in Tables 1 and 2.

In judgmental estimates, energy demand is normally broken down into three primary use sectors—residential and commercial, industrial, transportation—and one energy "transformation" sector—

---

[2] The elasticity of demand implied in these estimates is approximately −0.15 over the range of $7.00 to $11.00 per barrel, oil equivalent. (That is, demand changes by 0.15 percent when the price changes by 1 percent, rising when the price falls, and falling when the price rises.) Naturally, this is at best a rough approximation, and it cannot be assumed that this elasticity holds for prices outside this range.

electricity—which transforms primary fuel into electrical energy, which is then an input into the three primary sectors. Demand in each of the primary sectors for a particular energy source (including electricity) is affected by fuel price and other economic and demographic variables.

Demand in the residential and commercial sector includes the following end-uses: lighting, air conditioning, television sets, refrigerators, and small household appliances (these use electricity almost exclusively); cooking and dryers (using electricity and gas); and space and water heating (oil, gas, and electricity primarily). The total use of fossil fuels by this sector in 1970 was 14,000 trillion Btu. With 2,900 trillion Btu of electricity (net), the total consumption was 16,900 trillion Btu.

The use of energy by the industrial sector varies considerably among industries. Total energy consumption by this sector was about 23,300 trillion Btu in 1970, of which 2,300 trillion (net) is attributable to electricity. Four industries account for about half of total industrial energy expenditures: primary metals for 21.5 percent; petroleum and coal products for 15.4 percent; food and kindred products for 8.5 percent; and stone, clay, and glass for 8.3 percent. Knowledge of substitution possibilities among fuels for these and other industries is quite limited.

The transportation sector includes autos, buses, trains, subways, and so on. The primary energy source for this sector is gasoline, and there is little prospect for very much substitution among fuels in the short-run. As a result, anticipated gasoline supplies, at various prices, should be assigned to this sector first when making estimates of the future energy supply-demand balance. The total energy consumption by transportation was 10,800 trillion Btu in 1970, of which only a trivial amount was electricity. Of the total, about 8,100 trillion Btu was provided by gasoline.

Electricity demand will be limited to the capacity available—either by prices or controls. Thus, we look first at supply. Electricity is produced from coal, oil, gas, uranium, and hydroelectric power. To generate the 1.56 million gigawatt-hours produced in 1970, fuels were consumed in the following proportions: nuclear, 1.4 percent; hydroelectric, 16.2 percent; coal, 46.5 percent; oil, 11.6 percent; gas, 24.3 percent. (Note how little oil is used.) This translates into 322 million tons of coal, 325 million barrels of oil, and 3.89 trillion cubic feet of gas.

Projections of electricity supply for 1980 will probably be the most reliable of any energy forecasts. Given a five- to eight-year

Table 3

JUDGMENTAL DEMAND FORECASTS FOR 1980

(trillion Btu per year)

Fuel	Residential and Commercial Sector	Industrial Sector	Transportation Sector	Electric Utilities	Total
Coal	150	6,500	—	11,700	18,350
Petroleum	6,800	6,000	22,000	3,220	38,020
Natural gas	10,500	12,000	1,000	6,000	29,500
Electricity (net)	6,000	3,600	20	—	—
Nuclear	—	—	—	7,400	7,400
Hydroelectric	—	—	—	3,500	3,500
Total	23,450	28,100	23,020	31,820	96,800

planning and construction horizon, most of the additional supply that will be available in 1980 is from generating plants either under construction or in advanced planning stages.

The projections in Table 3 are a combination of available predictions of demand in 1980, as adjusted by our own credibility weightings. They are based largely on studies by National Economic Research Associates (NERA), Morrison and Readling, the National Petroleum Council, and Chase Manhattan Bank.[3] The studies that we relied on most heavily were those by NERA and NPC, both because they made fairly detailed projections for 1980 and because they carefully spelled out the derivations of their figures.

We first examined the total consumption forecasts for each of the four broad consumption categories. The variance for residential and commercial consumption forecasts was quite small for all studies, with a range of about 22,500 to 25,000 trillion Btu; we picked a value close to the mean. Industrial consumption forecasts were all fairly close, except for NERA's, which was far above the others. We therefore omitted the NERA figure and chose a value close to the highest

---

[3] The sources are the following: Chase Manhattan Bank, "Outlook for Energy in the United States to 1985," June 1972; Warren E. Morrison and Charles Readling, "An Energy Model of the U.S. Featuring Energy Balances of the Years 1947-1965 and Projections and Forecasts to the Year 1980 and 2000," U.S. Bureau of Mines, 1968; National Economic Research Associates, "Fuels for the Electric Utility Industry 1971-1985," August 1972; National Petroleum Council, *U.S. Energy Outlook: A Report of the Committee on U.S. Energy Outlook*, December 1972.

of the remaining three (the range here was between 26,400 and 28,500 trillion Btu) in order to give some weight to the NERA projection. The forecasts for transportation ranged from 21,700 to 25,700 trillion Btu; we took the mean. For electricity, the NERA forecasts were adopted because they specialize in this area and seem to be most familiar with construction and power-generation trends in the industry.

Predicting the consumption of specific fuels by each of the four sectors was more difficult. We used NERA's estimates of electricity consumption by each sector. NERA and NPC had virtually identical estimates for Btu conversion requirements based on fairly well-established engineering conversion coefficients, so the NERA data were used here as well. The NERA data were utilized to allocate total Btu requirements for electricity generation among fuels, because it seems that they had the best access to available data and appear to have analyzed it carefully.

Allocating our predicted 1980 demand among fuels in the transportation sector was no problem, since this sector consumes petroleum almost exclusively.

The residential and commercial sector uses only a slight amount of coal, and its proportion of total residential consumption has been declining. We projected a continuing linear decline to 1980, and, after subtracting electricity consumption, allocated the remaining Btu's between petroleum and natural gas on the basis of recent usage figures.

For the industrial sector, there was little to go on, since neither NPC nor NERA made the relevant projections for each fuel. We therefore took the 1970 proportions by fuel and applied them to the industrial demand left after subtracting electricity demand.

The resulting total energy demand is 96,800 trillion Btu in 1980, as shown in Table 3. Converted to oil-equivalent units, this is the demand for 45.6 million barrels per day that appears in Tables 1 and 2.

Demand Reduction by "Conservation"

The demand estimates in Tables 1 and 2, made using the Hudson-Jorgenson model, showed a reduction of 1980 demand from 44.2 million barrels per day at $7.00 per barrel to 40.6 million barrels per day at $11.00. This change is presumably the result of substitutions for energy, conservation of energy in consumption, and increased energy productivity in production of goods and services. An alternative way to estimate the potential for reductions in demand is to determine where new or existing technology can bring about reductions in energy consumption without degradation in function performed. When supplemented by appraisals of conservation efforts (based on experience in the last ten months), these estimates can serve as an indication of the potential for demand reduction induced by government policy and higher prices.

In the industrial sector, a rough survey of the responses to increasing fuel prices and fuel shortages shows that a 15 to 25 percent reduction in energy consumption can be obtained by eliminating heat leaks and improving waste-heat recovery. It is hard to know how many such changes, recently made, were caused by the threat of shortage and by public relations campaigns, as compared to rising prices. Thus we cannot know if further reductions might result from high fuel prices alone, or whether they would require special incentives (such as fast tax write-offs) or various forms of controls. Without attempting to specify the relative influences of price and other incentives, we will take the 15 to 25 percent potential savings in the industrial sector as a "once-and-for-all" demand reduction in industrial fuels between now and 1980.

Transportation fuel demand is predominantly determined by consumer attitudes, and therefore has a high component of uncertainty. Rather than attempting to predict changes in consumer tastes, we have considered a relatively simple set of feasible technological adjustments; they could be accomplished by modification of new cars and some retrofit of old cars. Electronic ignitions on all new cars and part of the existing stock could save by 1980 an amount of fuel equal to 5 to 10 percent of the consumption by all cars and a large proportion of trucks. This would come to an estimated 60 to 120 million barrels of gasoline in 1980. Radial tires on all new cars, plus replacements on old cars, would result in a 5 to 10 percent savings, yielding 45 to 90 million barrels of gasoline in 1980. Policies to discourage automatic transmissions and air conditioning, leading to a 50 percent reduction of these items during the next six years, would yield a 95 million barrel savings of gasoline. The total savings in gasoline from these changes would be from 200 to 305 million barrels in 1980.

Potential reductions in consumption in the residential-commercial sector have been estimated on the basis of possible improvements in the existing stock of buildings. Items considered included storm windows and doors, six-inch ceiling insulation, sealing and weather stripping, solar assist and energy storage in heating, heat pumps, heat exchangers, and increased burner efficiencies. These improvements were considered by fuel and by region of the country.

The measures we have discussed do not require a change in life style or drastic limitations on the use of energy. The experience of the last six months in New England indicates that a savings as great as 15 percent in heating oil is attainable by thermostat (and attitude) adjustments which do not change living conditions significantly. A similar adjustment has occurred in the consumption of electricity—in the short run, at least. The transportation sector was less responsive initially, but a combination of price increases and shortages has produced 10 to 15 percent short-term reductions.

These results make it reasonable to expect that a continuing and forceful campaign of public information could significantly influence energy-use patterns in the transportation, residential, and commercial sectors. Without direct controls, consumption could probably be reduced by 10 percent of 1971 consumption. Summing over all the estimates we have made, it is possible to show savings in a range from 4 to 8 million barrels per day by 1980.

This gross estimate may be compared with the demand estimate of 51.1 million barrels per day prepared by the Chase Manhattan Bank. The Chase forecast was made before problems of insecurity of supplies and rising price were evident, so notions of energy conservation were not considered. Correcting the Chase forecast by our estimate of savings yields a demand of about 43 to 47 million barrels per day, which brackets the judgmental forecast reported earlier in this section.

1980 and Beyond

The rough dimensions of energy supplies and demands in 1980 show that complete independence from imports of energy will probably cause prices of $11.00 to $13.00 per barrel, or more. Beyond 1980, however, the picture could change. We have not examined conditions in later years in sufficient detail to determine how supplies and demands might balance, for our purpose here is to evaluate the goal of self-sufficiency by 1980. It may be interesting, nonetheless, to briefly speculate about the possibilities for independence in 1985.

Our speculation is that in 1985 self-sufficiency can be achieved only at price levels (in 1973 prices) very similar to those shown in Tables 1 and 2 for 1980. For market-clearing prices to be lower than $11.00 to $13.00 per barrel, additions to domestic production would have to exceed additions to demands from population increase, economy-wide income increases, and other factors. According to the Hudson-Jorgenson model simulations, demand growth will be approximately 2.8 percent per year at prices in the range of $9.00 to $10.50 per barrel. At this growth rate, about 1.2 million barrels per day are added to demand each year—a total of 6 million barrels per day over the period 1980 to 1985.

Thus new supplies from new sources over this five-year period would have to exceed 6 million barrels per day—and exceed this figure by enough to compensate for depletion of domestic oil and gas, *and* bring prices down. Given the magnitude of the construction required to bring on new technologies at that supply level, and the consequent effects on costs of fuel production from these new technologies, such huge additions to supply are not likely to be forthcoming. A hopeful—if not overoptimistic—forecast is that additions to supply, working against inexorable demand increases, would be sufficient to maintain prices in the neighborhood of $10.00 to $11.00 per barrel under a program to achieve autarchy in 1985. Once again, though, there is great uncertainty inherent in such a prediction.

3
DOMESTIC PETROLEUM
AND NATURAL GAS SUPPLIES

The Petroleum Supply

The production of crude oil in 1973 was 9.2 million barrels per day. The total consumption was about 17.0 million barrels per day, which included about 6.1 million barrels per day of imports, and 1.7 million barrels per day of natural gas liquids. To raise production to higher levels by 1980 would require prodigious exploration and development, as an appraisal of present conditions and forecasts shows.

The National Petroleum Council's projection for 1980 expects Alaskan North Slope production to be from 2.0 to 2.8 million barrels per day, although 2.0 million barrels is likely to be closer to the truth due to delays in pipeline construction. Continental U.S. offshore production is expected to range between 1.6 and 2.7 million barrels per day. Total production of petroleum liquids is projected to be from 8.9 to 13.6 million barrels per day, depending on a number of political and economic factors. Thus the most hopeful NPC projection—13.6 million barrels per day in 1980—could require up to 8.9 million barrels produced onshore (in the lower forty-eight states) as compared to 7.8 million in 1972 and 7.6 million in 1973.[1]

In 1973, onshore production was 12 percent of onshore reserves. Assuming that this ratio can be maintained, and that there will be a smooth buildup to a 1980 production of 8.9 million barrels daily, 21.1 billion barrels will be consumed in seven years, and 27.1 billion barrels of reserves will be needed to support production in 1980. The

---

[1] In preparing these estimates, and others below, we have attempted to reconcile differences between detailed figures provided in the initial task group report of the oil subcommittee of the NPC study, and the ultimate summary report, *U.S. Energy Outlook*, December 1972.

total is 48.2 billion barrels; subtracting the current reserves of 22.8 billion barrels leaves 25.4 billion to be added in seven years, or 3.6 billion per year. This total has been equalled or surpassed only three times since World War II. It is technically feasible, but it must now be done under much less favorable conditions than before.

It is generally agreed that discovery of new fields has been dwindling for a long time. In 1946-1949, gross new reserves developed were 11.9 billion barrels. But 9.1 billion was provided from fields newly discovered during those years; hence discoveries made up for three-fourths of new reserves developed. This fraction dwindled steadily, and in 1965-1967, while 7.6 billion barrels of new reserves were developed, only 2.0 billion were added from newly discovered fields. (The percentage was even lower in later years, but data on recent-year discoveries are inherently incomplete and should not be used.) In other words, the industry has to an increasing extent created new reserves in old fields, both by finding new oil in extensions and new pools, and by improving recovery. This process should continue, but it is unreasonable to expect it to continue at the average of past costs. That is, the less our underground stock is replenished by *new* fields, the harder we must work the *old* fields. Thus the costs of developing new reserves and productive capacity will rise.

An indication of considerably increasing costs is provided by reserves added by the completion of an oil well. The average increased from a low of 80,000 barrels in 1957 to a peak of 218,000 in 1970, then fluctuated between 142,000 and 212,000. This development is ominous because of the decline in drilling and in new reserves. During 1955-1970, the number of wells drilled fell from 34,000 to 14,000 per year, but the amount of new reserves established declined only mildly. Because of better regulation, fewer useless wells were drilled, and higher capacity, lower-cost wells took a disproportionate share of the increase in demand in the late nineteen-sixties. Thus the rise in development costs due to a constantly less favorable resource base was offset by a once-for-all regulatory improvement and the resulting concentration on better prospects. Since the number of new oil wells decreased from 12,800 per year in 1970-1971 to 10,600 in 1972-1973, oil operators must have continued to drill even more selectively. For reserves added per well to shrink during such a time shows a rapid rise in the real cost of developing new reserves from old fields. If so, the past average of the cost required to develop reserves must be a serious underestimate of future costs needed to develop new reserves. As drilling again expands, the reserves added per unit of drilling must be expected to fall again.[2]

The Judgmental Forecast. Against this background, we consider the National Petroleum Council estimates for 1980. These are based on historical experience; they assume that what has been found and developed is a reasonably good sample of what will be found and developed. The NPC estimates embody the judgments of the best informed people in the industry, and at a disaggregated level, so that the judgments are relatively independent of each other. Also tending to make them plausible is the fact that unrecovered oil-in-place in the lower forty-eight states is about ten times proved reserves, which indicates that there is a big stock of oil which was uneconomical to recover at previous prices.[3]

The problem is that the NPC projections do not follow from a model, but are based almost entirely on perceived trends. The four principal supply cases that the NPC investigated were:

Case I: Expansive supply, requiring a vigorous effort fostered by early resolution of environmental controversies, ready availability of government land for energy resource development, adequate economic incentives, and a higher degree of success in locating currently undiscovered resources than in the recent past.

Case II: Less expansive supply, assuming improvement in finding rates for oil and gas, and a quicker solution to problems in fabricating and installing nuclear power plants than in the recent past.

---

[2] With depletion rates around 12 percent or more, costs would rise very rapidly if operators attempted to drain pools more rapidly. It can be shown that the average capital cost of any project is approximately equal to $(I/Q)\ (a+r)/365$, where I/Q is the investment needed per additional daily barrel, a is the decline rate, and r is the discount rate or cost of capital. That is, a price just equal to this amount would barely compensate for the investment. Increasing the rate of depletion would, of course, be less attractive the greater the rate of depletion already is, since it would require larger investment. It also can be shown that the speed-up cost is equal to the development cost multiplied by a factor $(a+r)/r$. Where a and r are approximately equal, as they are today, with both the rate of decline and rate of discount near or above 12 percent, the marginal speed-up cost is twice the incremental capital cost. For example, if it cost $5,000 to develop an additional daily barrel, and a and r both equalled 12 percent, the annual capital cost would be about $3.44 per barrel. But to speed up depletion of a given reserve would cost about $6.88 because, in effect, of the feedback on the existing operation. Hence the price would have to double before the previously break-even operation was worth expanding.

[3] Alaska deserves special mention. A very small amount of exploratory effort at the North Slope has yielded one gigantic find. Costs are very low by comparison wth current world prices, and the discovery of a few more Prudhoe-Bay-type reservoirs would change the national picture drastically.

Case III: Less expansive supply, assuming a slightly different but equally cautious mix of policy changes. The forecasts are much the same as those of Case II.

Case IV: Low supply, representing the likely outcome if environmental disputes continue to constrain the growth in output of all fuels, if government policies prove to be inhibiting, and if oil and gas exploratory success does not improve over recent results.

None of these cases incorporates price as a determinant of exploration or development. In the NPC analysis, "price" is the *result* of a given rate of drilling and finding. The NPC procedure is to assume alternatively a high, medium, or low drilling rate and a high or low "finding" or reserve-addition rate. From these are created the four cases: high drilling and finding rates in Case I, a medium drilling rate but a high finding rate in Case II, a medium drilling rate but a low finding rate in Case III, and low drilling and finding rates in Case IV. The apparent anomaly that Case II shows less produced at a lower price, while Case III shows less at a higher price, is thus not a mistake, but follows from the assumptions. In each case, the NPC has used averages of drilling and finding, but without trying to relate the two. Estimating the amount of capital spending needed to maintain a high drilling rate, for example, they take a 15 percent rate of return, and divide the required return in dollars by estimated production under a high finding rate, deriving a "price" of $5.64 per barrel (adjusted to the 1973 price level).

There are two reasons why this "price" which *results* from drilling and finding may seriously underestimate the price needed to *create* incentives for Case I supply. First, the NPC does not discuss the likely rate of expenditure for finding. The most likely combination is a very low finding (reserve addition) rate with a very high drilling rate. Onshore reserves added in 1971-1973 averaged less than 1.8 billion barrels per year, while drilling and real development expenditures did not decline correspondingly. If we are to double this performance, up to about 3.6 billion barrels per year, we can hardly expect to escape with less than three times recent outlays. Hence the NPC 1980 "price," which is only 54 percent higher in constant dollars than the 1970 price, cannot possibly be an accurate assessment of the necessary 1980 expenditures per unit of new reserves and capacity installed.

Furthermore, the NPC does not touch on the wide dispersion of onshore cost. If a given "price" is just enough to cover capital and operating cost of all new oil developments, this average will be greater than the costs of the lowest-cost capacity, but less than the highest. Only if the industry were to subsidize the production of high-cost oil out of the profits of low-cost oil could this "price" be treated as the price needed to bring in the high-cost oil. The range of cost is unknown, but must be very large.

Given these two problems with the NPC estimates—the most recent costs per barrel of new reserves added, and the need to cover high-cost sources—we can at least be certain that the price needed to bring in 13 million barrels per day in 1980, or possibly much more, must be far higher than the $5.64 estimated by the NPC. We might suppose that to double the 1972-1973 performance would require tripling the price, hence the $4.30 for old oil would increase three-fold to $12.90 in 1973 dollars. It seems unreasonable to assume a price below $9.00 if, with no imports, higher coal and gas production keep oil production to no more than 13 million barrels per day.

The forecasts shown for the NPC in Table 2 are a combination of Case I estimates and our judgment of the appropriate price for Case I. Under almost ideal conditions, the Case I production for the lower forty-eight states would be forthcoming at $7.00 per barrel (which is above the NPC's $5.64 but is not inconsistent with their report, since $5.64 is not the "supply price" equal to marginal costs). The Case I production is certainly consistent with $9.00 and $11.00 prices; since no additional supply is shown in the NPC study to be forthcoming at these high prices, we do not assume any more supplies than at the lower price.[4] It also is consistent with a continued decline in domestic output during 1974 despite a "new" oil price (see below, p. 76) of near $10.00 since late 1973.

The Econometric Forecast. The econometric forecasts of supply are derived from the MacAvoy-Pindyck econometric model.[5] The approach taken in the equation formulation of the model is to treat oil and gas exploration as an investment decision under uncertainty which results in oil and gas as joint products. Discoveries are the

---

[4] The exception is Alaskan oil, which is shown on a separate line in Tables 1 and 2. While the table shows 2 million barrels per day from Alaska at all prices, this amount must be more secure at $11.00 per barrel than at $7.00.

[5] The preliminary version of this model and its application to policy analysis is described in P. W. MacAvoy and R. S. Pindyck, "Alternative Regulatory Policies for Dealing with the Natural Gas Shortage," *Bell Journal of Economics and Management Science*, vol. 4, no. 2 (Autumn 1973). The final version is described in detail in P. W. MacAvoy and R. S. Pindyck, *An Econometric Analysis of the Natural Gas Shortage* (North Holland, forthcoming).

product of wells, the proportion of successful wells, and the size of hydrocarbon finds per successful well. Each of these three variables depends on prices of oil and gas, the cost of exploration, and the inventory of available but previously undiscovered deposits. Equations have been formulated and fitted to data for the 1960s and early 1970s for these relationships; these are used to forecast discoveries. Other equations are formulated for extensions and revisions of previous discoveries (without price effects) and for production out of total reserves (with no price effects for oil, but some price effects for gas).

The model shows that prices make some difference in discoveries, but that it takes a long time before there is any noticeable effect on production. Price changes over the range of $7.00 to $11.00 per barrel in real terms increase discoveries by approximately 25 percent (so that the elasticity of supply of discoveries is roughly 0.5 in that range). These changes in discoveries have a small effect on *total* reserves—increasing the forecast level of 1980 reserves from 20.4 billion barrels at $7.00 to 21.8 billion barrels at $11.00 per barrel. Since total reserves determine production, the effect is to increase domestic U.S. production from 6.5 million barrels per day at the lower price to 6.9 million barrels at the higher price.

The important determinant of both discoveries and production, as a matter of course, is the amount of oil reserves still waiting to be found. The "success ratio" and "size" equations show strong depletion in the continental U.S., so that finds in known regions decline in size over the rest of the decade. This is because that has been the historical pattern; the model "tracks" 1967-1972 reserve additions within 5 percent. The only reason this would not continue to be the case would be the opening of new regions offshore for which depletion would not have occurred. This is possible, if not probable; the fact that the model does not deal explicitly with offshore regions is a defect in forecasting that might lead to downward-biased forecasts.

Another reason for questioning the econometric forecast is that the data on which the equations were estimated are not entirely appropriate for that task. During the period over which the data were generated, the real prices of oil and gas did not change greatly. This creates two problems in using the equations. One is that of using the supply "elasticity" to forecast production response of a price change outside the range of all previous price changes (and thus outside the range of historical equation fits). Perhaps the responsiveness would be much greater for a large change than for the previous small changes. The other problem is whether the fitted

26

relationships between discoveries and prices are "reversible." In the past, discoveries and prices both declined so that equations fitted to the data showed a positive relationship. In the future—according to the alternative policies under review here—both discoveries and prices increase. The question is whether movements in both directions are the same.

The forecasts from the model are lower than the judgmental forecasts. They show stronger depletion and price effects than in the judgmental, so that both control variables seem important. But, because of the problems raised in the preceding paragraphs, we would not use these forecasts to the exclusion of judgmental projections. The best estimates probably lie between the two.

The Natural Gas Supply

The shortage of natural gas in the United States has grown rapidly in the last two or three years. By now, it probably exceeds 10 percent of total demands. This is not a result of a shifting of demands to natural gas due to the Arab oil embargo. Rather, it is a continuous and systematic long-term shortage, and there is every reason to believe that it will not be eased appreciably in the remaining years of this decade under the prevailing Federal Power Commission price controls—even if the FPC were to continue its recent policies of raising the price an average of three or four cents per thousand cubic feet on *new* contracts each year. If great pressure is put upon gas demands as a result of oil price increases beyond $6.00 per barrel, excess demand is likely to expand to more than a quarter of total demand in the next few years.

Because of the control mechanism operated by the FPC, production of gas could either increase or decrease greatly in the next few years. In order to provide a basis for analysis of public policies in gas, we attempt to forecast supply with the MacAvoy-Pindyck econometric oil and gas model referred to earlier, assuming two quite different policies regarding gas prices. Both are alternatives to the status quo.

Supply under Price Controls. The first alternative is to direct policies toward a price freeze. This is likely to occur under restrictive regulation such as that implied by Senate Commerce Committee bill S. 2506 (the "Stevenson bill"), which calls for an expansion of regulatory jurisdiction for the FPC to cover all sales at the wellhead (including intrastate sales). The bill requires that price ceilings be based upon historical average costs, so that this legislation seeks to stop the price increases now occurring under more "relaxed" FPC regulation. The price implication might well be to limit increases to approximately one cent per year when these increases are justified by changes in average costs of drilling and production. There are many possible variations on this interpretation, but it is unlikely that price increases much greater than three cents per year are implied by the bill, since price increases of five cents a year are now being put into effect by the FPC, and the bill specifically delineates standards which would not allow such increases. The general thrust of this and similar legislation is to hold the line on present prices so as to prevent gas sales from following the "pricing spiral" now seen in petroleum sales.

Holding the price line implies that there will be little additional incentive to explore and develop new reserves or to restrict demands for natural gas as the price of fuel oil increases. Under this policy, new-contract field prices are assumed to rise, on average, at a rate of one cent per year, from roughly 35 cents in 1973 to 58 cents in 1980. As a result, wholesale prices throughout the United States are expected to rise on average to 67.4 cents by 1980. Assuming that wholesale prices rise 6.5 percent per year, and population increases at one percent per year, the market for natural gas is expected to grow. Moreover, conditions in oil markets have strong implications for excess demand in natural gas markets. Rising oil prices lead to substantial increases in demands for natural gas which cannot be satisfied at the regulated or frozen level of prices.

The effects of strong price controls are shown in Table 4. The low levels of annual production and high levels of demands result in significant excess demands—2 trillion cubic feet at the present time, increasing to 11 trillion cubic feet in 1980, or approximately 25 percent of total 1980 demands. This shortage would be so great as to make it impossible for the pipelines to supply all the needs of established consumers. Permanent rationing would undoubtedly be put into effect. Most of its impact would be felt in the upper Midwest, where population and industrial growth are large and where the pipelines serving the region come from producing areas that are most depleted.

Supply under Price Decontrol. The second alternative to present policy is the elimination of current restrictions on field prices. The purpose would be to provide incentives for increasing reserves and production (by higher prices) and for eliminating low-value uses of

28

Table 4

NATURAL GAS SUPPLY UNDER A REGIME OF STRICT PRICE CONTROLS

Year	Field Prices on New Contracts (cents per thousand cubic feet)	Additions to Reserves (trillion cubic feet)	Production (trillion cubic feet)	Demand (trillion cubic feet)
1972	31.7	8.8	23.3	23.5
1973	34.7	17.5	23.7	24.3
1974	39.7	19.0	24.6	26.3
1975	42.8	21.7	25.3	28.8
1976	45.9	23.6	26.2	31.3
1977	48.0	25.4	27.0	33.7
1978	52.0	26.6	28.0	36.2
1979	55.1	26.2	29.1	38.6
1980	58.2	24.5	30.2	41.1

natural gas (by reducing demands through price increases). The use of market forces to add to supplies and reduce demands would have different effects, depending on how rapidly and extensively prices increased. Immediate and complete elimination of price controls would establish short-term equilibrium prices much greater than those that would persist over the long run—particularly if there are short-term shortages in alternative fuels such as fuel oil. Phased decontrol, however, could be put into effect in a way that would gradually ease restrictions on prices, so as to move them over a five-year period into long-run equilibrium.

Thus this policy does not require immediate deregulation: new-contract price increases could be limited by the Federal Energy Agency for some years, presumably to keep the increases in line with general cost-of-living increases. This implies a ceiling on new-contract prices of approximately sixty-five cents in 1974 (the average price was thirty-five cents in 1973). There would be a five-cent price increase each year thereafter. Field price increases would feed through as price increases charged by pipelines to wholesale buyers, so that the immediate impact would be an eight-cent increase in wholesale gas prices across the country. By 1980, field prices on new contracts would

Table 5

NATURAL GAS SUPPLY UNDER PHASED PRICE DECONTROL

Year	Field Prices on New Contracts (cents per thousand cubic feet)	Additions to Reserves (trillion cubic feet)	Production (trillion cubic feet)	Demand (trillion cubic feet)
1972	31.7	8.8	23.3	23.5
1973	34.7	17.5	23.7	24.3
1974	39.7	19.0	24.6	26.4
1975	64.7	25.6	26.8	28.7
1976	69.7	31.3	28.2	30.5
1977	74.8	35.8	29.2	32.0
1978	80.0	41.5	31.0	33.2
1979	85.1	42.8	33.0	34.3
1980	90.3	41.8	35.0	35.2

rise to more than seventy-three cents, while wholesale prices on all contracts would average eighty-four cents.[6]

The price increases would substantially increase additions to reserves over a five-year period, and they would increase production, both because of the reserve additions and because of more intensive depletion of existing reserves. As shown in Table 5, production is expected to rise from 26 trillion to 35 trillion cubic feet over the period 1974-1980. At the same time, demands would slow down as consumers attempt to avoid the price increases. Excess demand is 1.8 trillion cubic feet per year in 1974, grows to 1.9 trillion cubic feet in 1975, but then declines to 0.2 by 1980. In effect, the use of market forces should eliminate excess demand through a combination of additional supplies, reductions in use by buyers faced with higher fuel costs, and substitutions for this higher priced fuel.

The M.I.T model simulations therefore support the position that phased price increases (leading to a reliance on market forces over the long run) can be used to ameliorate the present and growing shortage

---

[6] Currently, contracts for *intra*-state gas are reportedly selling for as much as seventy-five cents to $1.00 per thousand cubic feet. This appears to imply a free-market price higher than we forecast. Actually, there need be no conflict here, for there is great excess demand in intra-state markets (some is demand diverted from other states) which is impinging on a narrow market for new gas. With phased deregulation of prices, this pressure would be relieved.

of natural gas. As price incentives improve the profitability of additional drilling in the United States, more reserves are likely to be accumulated, and more production can then take place. Most geological estimates indicate that these reserves are available, albeit at higher costs of discovery and extraction. Higher prices to consumers will be more representative of the true value of this scarce resource, and consumer demand should respond accordingly.

Supply Forecasts in Tables 1 and 2. The supplies of gas likely to be forthcoming at different levels of oil prices can be estimated as follows: First, it is assumed that gas prices will be regulated for the rest of the decade, either according to "strong price controls" as in Table 4, or in the process of "phased deregulation" of prices, as in Table 5. Either way, markets will not be allowed to operate so as to eliminate the present differences between gas and crude oil prices. Second, it is assumed that crude oil prices of $7.00, $9.00, and $11.00 per barrel enter gas markets as an "outside" variable on the demand side—that is, higher oil prices add to gas demand—and on the supply side—higher oil prices add to discovery and production of gas, which is often found with oil. Inserting into the M.I.T. model the gas prices in Tables 4 and 5, oil prices of $7.00, $9.00, and $11.00 result in the supply forecasts shown in Table 6.

The table shows that gas markets clear by 1980 under phased decontrol over part of the range of oil prices. Higher oil prices add

---

Table 6
THE SUPPLY OF NATURAL GAS AT
VARIOUS PRICES PER BARREL OF OIL
(trillion cubic feet)

Year	Gas Supply under Strict Price Controls, with Crude Oil at:			Gas Supply under Phased Price Decontrols, with Crude Oil at:		
	$7.00	$9.00	$11.00	$7.00	$9.00	$11.00
1976	26.2	26.2	26.2	28.2	28.2	28.2
1977	27.0	**27.0**	**27.0**	29.2	29.2	29.2
1978	28.0	27.9	27.8	30.9	30.9	30.8
1979	**29.0**	28.9	28.7	**32.9**	32.8	32.6
1980	**30.1**	**29.8**	**29.6**	**35.0**	34.6	34.3

Note: On the left half of the table, numbers in bold face type indicate excess demand greater than 10 trillion cubic feet. On the right half, the bold face numbers indicate excess demand less than 1 trillion cubic feet.

more to gas demand than they do to gas supply; at $7.00 oil prices the net additions to gas demand are not substantial, but at $11.00 they add enough to make excess demand for gas 10 percent of total demand. For the same reason, gas price controls similar to "strict regulation" are much worse under high oil prices than under low oil prices; the size of the shortage exceeds 10 trillion cubic feet as early as 1977 if oil prices are pushed to $11.00 per barrel in 1975.

The estimates for 1980 given in Table 6 have been inserted in Table 1 (in millions of barrels per day). The imposition of strict regulation is taken to be the most likely policy over the decade, even if Congress does not pass the Stevenson bill. After all, this has been the most systematic long-term policy of the FPC. Thus the forecasts on the left of Table 6 are used in Table 1.[7]

Much the same approach has been used in selecting the National Petroleum Council forecasts for gas liquids and for natural gas shown in Table 2. Here the Case II forecasts are shown, rather than the Case I as in crude oil, because gas price controls are likely to exert a significant influence over the next few years. Even with partial decontrol, the time lags in exploration prevent additions to supply before 1980 which are much more expensive than those of Case II.

---

[7] Note that, over the range from $7.00 to $11.00 per barrel, the supply of gas declines. This is because the oil price goes up while the gas price continues to be controlled close to fifty cents per million Btu, which makes gas exploration less attractive relative to oil exploration, so that gas reserves and production decline slightly.

4
THE COAL SUPPLY

The United States has vast reserves of coal, both in the well-explored regions east of the Mississippi River and in partially explored portions of Wyoming and Montana. Whether these reserves can be developed in time to make a significant contribution to Project Independence depends upon the cost of mining and transporting the coal to final markets and on the growth in demand for coal. These factors, in turn, are influenced by political matters—principally environmental protection regulations against sulfur emissions from coal burned to generate electricity and strip-mining legislation.

The potential pattern of development calls for operating on the *intensive* margin in the eastern coal regions: exploiting existing underground mines more intensively, first for sulfur-free coal, and later for sulfurous coal. Before that sulfurous coal is considered, however, above-ground strip mining of new resources in Wyoming and Montana should come into operation on the *extensive* margin, thereby helping avoid violation of environmental standards. Thus the potential supply of coal at various prices consists of eastern low-sulfur sources at now prevailing prices, western strip-mined coal at prevailing or slightly higher prices, and eastern high-sulfur coal at much higher prices—where the higher prices include the implied social costs of environmental degradation, or, alternatively, the costs incurred for stack-gas scrubbers or other purification devices.

There is a great deal of high-sulfur coal available at these higher prices, but the ability of the economy to utilize it is limited. This is because only so much coal can be used for energy—coal cannot be burned in automobiles or airplanes—and in the next seven to ten years the capacity of facilities able to burn coal is relatively fixed.

In an attempt to quantify this state of affairs, we begin with estimates of the unit cost of coal at present levels of production in the East, and at now-contemplated levels of production from the western strip-mined regions. We will then ask whether costs of extraction would be significantly greater at higher rates of production, and thus begin to trace out a rough supply function for higher prices. At that point, the supply function will be "truncated" by introducing demand constraints.

The Cost of Eastern Low-Sulfur Coal

Because of the severe environmental damage caused by the technique in the hilly terrain of Appalachia and the high cost of reclamation, it is likely that strip mining will be limited there. In the flatter sections of Illinois and Indiana, there is evidence that large blocks of strippable coal are scarce. Therefore, any substantial expansion of coal production in the East will have to rely on underground mining.

The unit costs of coal extraction in the East are a composite of capital costs, labor costs, and developments in labor productivity. Concentrating on the low-sulfur coal, capital costs appear to be close to $2.50 per ton, a figure based on recent information for capital expenditures and a 15 percent rate of discount over a twenty-year period.[1] Supplies, power, and other minor inputs, according to Bureau of Mines engineering estimates, add about $2.00 per ton.

Labor costs depend upon assumptions about wage increases and productivity changes. The projection of wages is a difficult problem. Wages are determined through collective bargaining and there is no unique wage for each level of employment and output. Therefore, at current levels of output a wide range of wages is possible. To any estimate must be added about an eighty cents per ton contribution to the union welfare fund. R. L. Gordon estimates that in 1973 the average daily wage, including fringe benefits, was $65.60.[2]

This wage is not likely to remain constant, even at constant levels of employment. It is becoming increasingly difficult to attract new workers into coal mining. Moreover, the entry of inexperienced workers necessitates training costs not incurred when mining companies were able to draw upon an experienced labor pool. For these

---

[1] In February 1973, the *Mining Congress Journal* reported the opening of a new metallurgical mine in the East at a cost of about $16.00 per annual ton, which is probably on the high side since it was for metallurgical-quality output. This was annualized at a 15 percent discount rate over twenty years.

[2] See Richard L. Gordon, *The Competitive Setting of the U.S. Coal Industry, 1946-1980* (forthcoming).

34

reasons labor costs will rise, although it is impossible to say how high. An extrapolation of the 1969-1973 rate of real wage increase yields a $97.00 daily wage in 1980. Allowing for a truly extraordinary increase, we take $150 as an upper limit by the end of the decade.

Labor productivity has not been increasing, and it is tempting to assume constant or declining productivity to 1980. This might be unduly pessimistic. The industry is beset by problems that, hopefully, are transitional. The Health and Safety Act of 1969 introduced many changes in mining procedures that are still having an effect; but if these are adjusted to, productivity should return to its 1969 level by 1980. We take as a 1969 productivity level that of a large new underground mine producing about twenty tons per man-day. Together with a daily wage of $150, this yields total 1980 costs of fifty-three cents per million Btu. A more optimistic productivity figure of twenty-five tons per man-day yields a cost of forty-seven cents for the same quantity.

These estimates lead to a point estimate of "supply." The 1980 total of capital, materials, and labor costs is roughly $3.80 to $4.20 per barrel, oil equivalent, delivered in Detroit. It seems reasonable to assume that the supply of eastern low-sulfur coal (less than 1 percent sulfur) will remain near its present level of approximately 200 million tons per year at this price. At higher prices, additional supplies should be forthcoming. For example, we assume that at $7.00 per barrel, the production of low-sulfur eastern mines might increase to 250 million tons by 1980. Either estimate is a cautious extrapolation of present conditions.[3]

The Cost of Western Coal

Recent engineering studies of the Bureau of Mines establish the cost of mining coal in the Powder River Basin of northeastern Wyoming

[3] The evidence on reserves of low-sulfur eastern coal is very poor. Conventional estimates grossly overstate availability by including all coal in the ground, regardless of cost of extraction. On the other hand, a 1967 Bureau of Mines survey, *Analysis of the Availability of Bituminous Coal in Appalachia, 1971,* cited only 4.5 billion tons of recoverable reserves with less than 1 percent sulfur being held by producers of more than 100,000 tons per year. Assuming a mine life of twenty years, this could support production of only 225 million tons per year, or 25 million more than at present. The Bureau of Mines figure is undoubtedly an underestimate, since it excludes reserves held by land companies and smaller producers. In addition, what is considered recoverable would increase as prices reached historically high levels. Yet allowance must be made for replacing reserves lost through depletion. The recent difficulty utilities have had in obtaining low-sulfur coal in the East is further evidence of the inelasticity of supply of low-sulfur eastern coal.

and southeastern Montana.[4] The region is a large new source of coal. Costs there are approximately $2.25 per ton exclusive of royalties and state taxes, and assuming discount rates of 15 percent per year. (We exclude royalties since these are the returns that owners of low-cost or nonmarginal reserves would earn. State taxes have been excluded because, while they vary from state to state, at present they are negligible on the whole. The highest tax is in Montana, which adds thirty-four cents per ton or two cents per million Btu.)

These costs are low. Assuming 17 million Btu per ton, the Bureau of Mines figures yield a cost of about 13.2 cents per million Btu.[5] But these estimates may not include further social costs. The Bureau of Mines used the 1969 costs of land reclamation in their estimates, yet standards in many states have become stricter since then and will become even more exacting with the passage of a new strip-mining law. The costs depend on the amount of overburden that must be removed per ton of coal uncovered and on the topographical and climatic conditions of the area. Assuming reclamation costs of $5,000 per acre and a coal seam thickness of ten feet, then costs of reclamation are twenty-eight cents per ton or 1.6 cents per million Btu at the most.[6] This meets the environmental protection requirements at the present time, but this may not be enough to restore mined land to its former usefulness in agriculture. No one knows what the cost of complete restoration would be.

These estimates of costs reflect present conditions. Three factors could change future costs: depletion of coal reserves, changes in wages, and changes of transportation rates for coal from Wyoming to the Midwest. In predicting the effect of depletion on cost, the important factor is the "overburden ratio"—the thickness of the removed layers of rock and soil compared to the thickness of the seam of coal that their removal exposes. Bureau of Mines estimates of strippable reserves are based, at least for western coal, on main-

---

[4] Bureau of Mines, "Cost Analysis of Model Mines for Strip Mining for Coal in the United States," Information Circular 8535, 1972.

[5] A check on this estimate is the eleven to twelve cents estimated in 1970 by the North Central Power Study. (See *North Central Power Study Report on Phase I,* October 1971.) Allowing for approximately 19 percent inflation in construction costs since 1970, the *North Central Power Study* figure becomes thirteen to fourteen cents per million Btu in 1973 prices. Recently announced contracts have been in this price range, an indication of negligible royalties and taxes.

[6] The highest published estimate appears to be the $4,000 to $5,600 per acre cited in *Final Environmental Statement, Proposed Plan of Mining and Reclamation for the Big Sky Mine,* Peabody Coal Co., Coal Lease M15965, Coalstrip, Montana, p. XII-28.

taining the present overburden ratio.[7] Thus, the bureau estimates that there are about 13.6 billion tons of sub-bituminous coal in the Powder River Basin that could be extracted by strip-mining techniques at overburden ratios no greater than is now the case in this area. If we assume that the life of a mine is twenty years, this represents an output of 680 million tons per year before depletion makes it necessary to mine at higher overburden ratios, and thus at increased costs.[8]

Labor costs can also increase, resulting in higher costs for western coal. To exploit western resources, workers must be attracted to underpopulated regions of the country. We have not estimated how high wages would have to rise to bring in sufficient labor, but we can examine the effects of higher wages. Using engineering estimates of costs, we separate the total into labor and labor-related costs and capital costs. The former accounted for thirty-four cents per ton in 1973. The union welfare contribution added seventy-five cents per ton. (It is expected to increase to eighty cents per ton mined in 1974.) If all else stays constant in real terms (constant level of productivity and capital costs), we can examine the effect of an increase in real wages by varying the rate of increase of wages and looking at the final cost in cents per million Btu. This is done in Table 7, which shows that wages do not greatly affect long-run costs in western mining.

Transportation rates are the third important determinant of the cost of Western coal. In some sections of the country water transport is an alternative to railroads. Yet expanded use of the waterways will increase congestion at the locks and increase costs. Because of this, and because such transport is available only to favorably situated plants, this analysis will be restricted to rail transport.

Rail rates are set by the few railroads that run into the region from the Midwest. Existing rates range between 5 and 7.5 mills (tenths of a cent) per ton-mile. Many utilities planning new coal-fired plants are using rates close to 5 mills to estimate their shipping costs.

---

[7] See Bureau of Mines, "Strippable Reserves of Bituminous Coal and Lignite in the United States," Information Circular 8531 (1971).

[8] An additional uncertainty with respect to western reserves is whether the low-sulfur supplies from this region satisfy the environmental standards of the Clean Air Act. The low-Btu content of the coal requires that it have a lower sulfur content than coal with a higher heating value, such as that produced in the East. Because of this uncertainty and other short-run unknowns to be discussed later, we have attempted to be conservative in our estimates of how far coal production in the West can be expanded.

Table 7

WESTERN COAL COSTS UNDER ALTERNATIVE ASSUMPTIONS ABOUT THE ANNUAL RATE OF WAGE INCREASES

(dollars per ton)

	1973	1980, with 2 Percent Annual Wage Increase	1980, with 5 Percent Annual Wage Increase
Capital costs	$ .51	$ .51	$ .51
Supplies	.52	.52	.52
Welfare contribution	.75	.80	.80
Labor and labor-related	.34	.38	.45
Insurance and other	.11	.11	.12
Additional reclamation	.28	.28	.28
Total cost per ton	$2.51	$2.60	$2.68
Total cost per million Btu	.148	.153	.158

Note: All costs were adjusted to reflect the rate of inflation. "Capital costs" were calculated by computing the present discounted value of purchased equipment at a 15 percent rate of interest. Finally, "labor-related" expenditures refer to items that the Bureau of Mines estimates as a percentage of other expenditures; thus these costs would increase as wages rise.

Source: "Cost Analysis of Model Mines for Strip Mining of Coal in the United States," Bureau of Mines Information Circular 8535, 1972, pp. 85-100.

There are reasons for arguing that, for a given customer, rates could be either in the range of five mills or in the range of seven to eight mills per ton-mile. It is possible, for example, that new coal-fired plants might initially be charged reduced rates because new coal consumers could move their planned facilities and use another transporter. This bargaining advantage is constrained, however, because a single railroad provides service to most of the coal-producing territory and is assured of a preponderance of the outbound rail movement. Although a lower range of rates per ton-mile might prevail for new mining capacity in the next five years, pressures for higher rates to offset increased wage and fuel costs might be expected to nullify the benefits of low "incentive" rates, especially on long hauls to the Midwest. It is also possible that the railroads will be able to restrict the supply of transportation services through the medium of the Interstate Commerce Commission rate-setting practices. This is particularly so with eastern and midwestern railroads, which have a strong interest in protecting the competitiveness of eastern and midwestern coal mines in face of competition from the Far West. Under these conditions, the rates may be closer to 7.5 than to 5.5 mills per ton-mile.

In Table 8 we consider these alternatives by presenting the delivered cost for coal in various cities, using a fifteen-cent mine-mouth production cost and two different transportation rates. The table indicates the importance of the transportation rate in determining the supply price of coal. It must be expected that under present procedures for setting rates, the higher rates would be in effect and therefore the higher supply prices of coal would prevail.

These considerations lead to a rough but fairly comprehensive picture in which supply prices and levels of production are not greatly different from those prevailing at the present time. The constraints on increased production in the West lie not in a lack of low-cost reserves, but rather in the supplies of input factors between now and 1980. We examine these bottlenecks in the following paragraphs.

Table 8
THE COST OF WESTERN COAL
DELIVERED TO VARIOUS PLACES IN 1980

	Chicago	Detroit	East Texas	Philadelphia
Railroad Miles	1,000	1,370	1,500	2,000
Mine-mouth cost per million Btu	$ .15	$ .15	$ .15	$ .15
Transport cost at 5.5 mills per ton-mile	.37	.45	.50	.66
Transport cost at 7.5 mills per ton-mile	.50	.61	.67	.89
Delivered cost at 5.5 mills per ton-mile	.51	.59	.64	.80
Delivered cost at 7.5 mills per ton-mile	.64	.75	.81	1.03

Cost-Increasing Factors in the Short Run

The preceding discussion is based on attempts to find "point estimates" of marginal and average costs of producing coal at normal levels of growth of the industry and under expected conditions in input factor markets. This provides an approximation of the supply function for coal—the curve of supply versus price—at levels of "production" in keeping with low rates of expansion of present capacity. To find other points on the supply function involves considering higher levels of production. The marginal and average costs of providing higher rates of production should themselves be higher.

There are a number of potential bottlenecks in the supplies of input factors which could make the cost of providing more production appreciably higher than the estimates shown for normal-growth production. These bottlenecks appear in transportation, mining machinery, environmental and land-use legislation, and manpower.

Potential transportation bottlenecks include limitations on the supply of hopper cars and track necessary to haul large amounts of western coal into the Midwest. It is likely that by 1980 enough cars could be produced, and enough immediate policy changes could be made, to allow a larger outflow of coal traffic. The measures—such as upgrading roadbeds, improving signaling, adding to siding capacity, and so on—could be undertaken by 1980, and probably would not appreciably increase shipping costs per ton. There are, of course, limits to the process, although no one knows at present what these limits are; prevailing opinion seems to be that with no appreciable increase in cost the traffic could be expanded to more than 200 million tons per year.[9] The willingness of the railroads to make the necessary investments for handling higher traffic volumes would depend on their assessment of its duration and profitability.

A potentially serious bottleneck for the western coal industry is the availability of mining machinery. Industry representatives indicate that capacity could expand to support a billion tons per year of strip-mined coal by 1985, but rarely does anyone ask what the costs of capital goods for strip-mining operations at this rate of production would be. Yet even at the present time, the industry produces enough machinery each year to add 15 million tons per year to production

---

[9] This assessment depends upon railroad transportation conditions. There are other transport modes that could be used, such as barging from St. Louis down the Mississippi and up the Ohio River. Yet significant increases in barge traffic would probably congest the locks and increase costs. Slurry pipelining would appear to be limited because of the constraint on water supplies in the arid West.

capacity. Thus without any scale-up, the mining-machinery industry would provide enough for 100 millions tons additional yearly output by 1980. Therefore it is likely that more production could be achieved without much higher costs of capital equipment.

There are potential public policy bottlenecks as well, such as regulations on reclamation of strip-mined land. A strip-mining bill now under consideration in Congress could increase costs from the level of our estimates. More importantly, there might be absolute prohibitions on mining in certain zones. In many areas it would be very difficult to restore mined lands completely because of the dry climate, although there have been some successes on an experimental basis. If the law requires complete restoration, or forbids mining where it is doubtful this can be achieved, then the reserve base will be reduced.

Manpower limits have been discussed above, particularly with respect to mining in the West. If it becomes more difficult than expected to bring men and the necessary support facilities such as living quarters, services, et cetera, to an underpopulated area, then outputs could be curtailed or costs could increase at an unchanged rate of output. In all, the combination of these bottlenecks and limits on resources could substantially increase the prices necessary to bring forth supplies beyond the amounts forecast above.

Based on the cost estimates made earlier, and considering the various bottlenecks, there should be somewhere near 150 million tons per year of western coal available by 1980 at prices equivalent to seventy-five cents per million Btu delivered in Detroit. Assuming that a doubling of costs results from doubling output, then at prices equivalent to $1.50 per billion Btu in Detroit, there could be supplies from western sources totaling as much as 300 million tons per year.

Supplies of High-Sulfur Eastern Coal

Based on our rough estimates for eastern and western supplies, the total supply of low-sulfur coal in 1980 could rise as high as 550 million tons per year, which is about equal to current production of high- and low-sulfur coal combined. This suggests that complete reliance on low-sulfur coal may not be possible within the range of prices cited above, and uncertainty about low-sulfur supplies in the East makes it unlikely that government policy will rely on low-sulfur coal alone. High-sulfur eastern coal will be used, which will involve higher costs—either the implied costs of environmental damage or the costs of installing stack-gas desulfurization devices. If such devices are not available in 1980, a policy choice will have to be made between the use of high-sulfur coal or very much higher fuel prices.

Reserve statistics indicate the availability of large quantities of high-sulfur coal in the East. These could be available at the *low* end of the price range cited above for low-sulfur eastern coal, because producers would not have to dig as deeply or mine coal seams as thin as for the low-sulfur coal. At seventy-five cents per million Btu delivered in Detroit, it is not too optimistic to assume that at least the present rate of 350 million tons per year could be maintained. But this price does not include the social costs of pollution or the costs of sulfur removal. At the conclusion of this chapter, we will consider supplies of high-sulfur coal at various costs of sulfur removal.

Demand Constraints in the Short Run

Even if the United States's coal industry could expand its production without limit, the country would be limited in its capacity to use coal in 1980. Based on projections of past consumption, the 1980 demand for coal will be about 700 million tons. Allowing for the greatest possible extent of conversion of electric power plants from oil and gas to coal, demand could rise another 75 million tons by that year. This assumes that 44 percent of oil plants and 12 percent of multi-fuel plants could convert to coal.[10] While the calculation is a rough one, it appears that the domestic demands for coal in 1980, based upon capacity to utilize it and upon present technologies, probably would not exceed 800 million tons per year. This amount is not far greater than the volume expected to be forthcoming in the "point estimates" made above, which assumed normal growth. Thus, although appreciable cost increases could be expected at much higher rates of output, these higher rates may be irrelevant as a result of demand constraints.

Supply Forecast in Tables 1 and 2

Given all the uncertainties, it is difficult to say with any precision what the supplies of coal will be in 1980. Here we present a basic estimate for conditions of "modest growth" in the coal industry. An attempt is made to present a range of estimates for alternative prices, but only within the demand constraints.

---

[10] Federal Power Commission, *The Potential for Conversion of Oil-fired and Gas-fired Electric Generating Units to Use of Coal*, 6 November 1973.

Detroit is taken as a reference market, and it is assumed that coal competes with crude oil there at an equal price per million Btu of energy. Therefore coal priced at seventy-five cents per million Btu is assumed to be equivalent to oil at $4.50 to $5.00 per barrel. Coal at $1.50 per million Btu is equivalent to oil at $9.00 to $10.00 per barrel. Tables 1 and 2 contain an estimate based on the assumption that some high-sulfur coal is burned. In order to approximate the full costs of using this fuel, it is also assumed that stack-gas desulfurization devices become available.

At present, there is much dispute about whether this technology can be forecast, and what it would finally cost if it were available. The range of estimates is wide, depending upon who makes them and whether the forecaster is talking about existing plants or building new ones. The forecasted costs for fitting existing plants seem to range from thirty cents to eighty-five cents per million Btu. (At eighty-five cents this technique could not compete with low-sulfur western coal today.) Tables 1 and 2 assume that the technology is available at the midpoint of the range—fifty-eight cents per million Btu.

At $7.00 per barrel oil equivalent, eastern high-sulfur coal could cost no more than forty-one cents per million Btu at the mine-mouth. This is a low price compared to the estimates above, so 1980 supplies of eastern high-sulfur coal are estimated at 250 million tons, a reduction from the present level of output. Eastern low-sulfur supplies are estimated at 250 million tons, and western coal at 200 million tons. If exports reach the 86 million annual tons predicted by the Bureau of Mines, this would imply about 610 million tons consumed domestically. This represents a modest increase in domestic consumption from the 1973 level, yet at higher costs imposed by expenses for sulfur removal. The total is the equivalent of 6.1 million barrels per day of oil, as shown in Tables 1 and 2.

With stack-gas desulfurization available at fifty-eight cents, the earlier cost estimates indicate that at a maximum of $7.68 per barrel ($1.28 per million Btu), substantial amounts of high-sulfur coal could be available.[11] It is not unreasonable to assume that at this price, 300 to 350 million tons would be forthcoming. If prices were to reach $9.00 per barrel equivalent, further supply increases from the West and East would likely surpass the demand-constrained level of 800 million tons consumed domestically (approximately 8 million

---

[11] At a wage of $100 per day, in place of the $150 assumed earlier, this coal would become available at about $7.00.

barrels per day oil equivalent as shown in Tables 1 and 2) plus 86 million tons for export.

It is important to stress that this result depends upon the availability of desulfurization technology at fifty-eight cents per million Btu. The market-clearing price could be higher if such devices prove more costly. If they are not available at all, it is likely that sulfur restrictions would not be met, which would involve a social cost that might be reflected in the price, depending on government policy.

These estimates are rough, but a central conclusion emerges. Without low-cost desulfurization techniques, reliance on low-sulfur coal could be very expensive. If, on the other hand, a low-cost desulfurization method is available, coal use can expand quite rapidly—subject only to limits on demand.

The longer the time period, the more the bottlenecks can be overcome. In the short run, we move along a supply curve that reflects fixed capacity in mining equipment and rail transportation. But in time, adjustments can be made in these capacities. As we move into the nineteen-eighties, we would expect the output levels discussed in this section to be available at lower costs, and larger outputs to be available at approximately the same costs. The final levels of supply and demand depend crucially on environmental goals in public policy and the ability of the coal industry to supply an environmentally acceptable fuel.

5
NUCLEAR POWER AND SYNTHETIC FUELS

Nuclear Power

The large nuclear reactors presently being built for electric power generation are known as "thermal" reactors, since the fission of uranium or plutonium atoms within them is caused by neutrons which are moderated in energy so as to be nearly in thermal equilibrium with their environment. Water reactors can use either heavy water (deuterium oxide, D_2O) or light water (H_2O) for thermalizing neutrons and removing the energy released by fission; the most prominent U.S. design is the light-water reactor (LWR). An alternative design uses graphite for moderation, and helium gas for a coolant, and is called the high-temperature gas-cooled reactor (HTGR). Both LWRs and HTGRs are in use or under construction in the U.S.

Development work is underway on other reactor concepts as well. The liquid-metal-cooled fast breeder reactor (LMFBR) is considered by many experts to be the most attractive future design. The system will produce a shower of neutrons sufficient to convert a nonfissionable isotope of uranium into a fissionable isotope of plutonium, as well as causing the fission that ultimately generates electricity. It appears likely that an LMFBR can be built which will produce more fissionable material than it consumes—hence the name "breeder." The LMFBR has been developed to the point of operation of several demonstration reactors in the U.S. and abroad, and efforts around the world are increasingly focusing on this important concept. More advanced breeder concepts include the gas-cooled fast breeder (GCFBR), the light-water breeder (LWBR), and the molten-salt breeder (MSBR). These systems are in an early research phase, and offer little likelihood of having any impact before the LMFBR program.

45

Light-Water Reactors. Figure 1 illustrates the present planning schedule for the initiation, design, construction, and start-up of a light-water nuclear plant. Clearly no plant orders in 1974 can influence the availability of electric energy in 1980 unless the lead time of ten years is reduced by 40 percent or more. The principal bottlenecks can be grouped into four categories: initial site analysis, preliminary regulatory matters, construction, and final regulatory matters.

Site analysis. The one-and-a-half year process of site analysis and selection cannot be reduced significantly in the next one or two years. Beyond that, however, site analysis delays could be reduced by planning "energy parks," in which a large amount of capacity would be constructed at a single location. The use of platform-mounted offshore nuclear plants might also ease the site problem somewhat. Orders for several offshore plants have already been placed.

Preliminary regulatory matters. Before construction of a new plant can begin, an Environmental Report and Preliminary Safety Analysis must be prepared, and the plant must pass a review by the Atomic Energy Commission (AEC). All this takes time, for the environmental and safety studies are presently unique to each plant. Generic review of standardized plant types is being advocated by the nuclear industry. With such a change, and with the preparation of a general environmental report for a single site to be used by many plants, this part of the time scale could be reduced by 50 percent or more.

A continuing complication is that any public intervention can presently cause delays of up to a year in hearings on the issuance of a construction permit. The AEC has not been successful, thus far, in streamlining hearings. The problem is under study, but no public date for presenting new guidelines has yet been announced.

Construction. The current manpower situation is critical. Labor productivity is low due to lack of experience, uniqueness of each plant, and the need for exacting quality control. Most architectural and engineering firms with the capacity to handle nuclear design are severely taxed. All these factors contribute to construction delays.

There is some hope for improvement. Duke Power does its own construction and claims a four-year construction period—one full year less than is shown in Figure 1. Similarly, European and Japanese experience indicates that four-year construction is possible without excessive cost to the utility. Moreover, by standardization of plant design and construction, the experience gained in building a first

Figure 1

PLANNING SCHEDULE FOR A LIGHT-WATER REACTOR PLANT

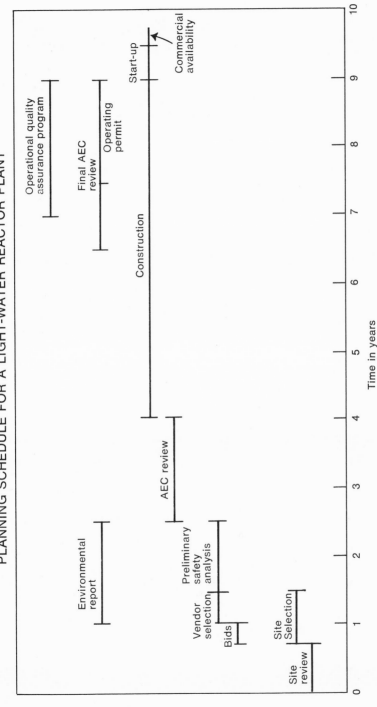

plant can increase efficiency and lessen construction time for subsequent plants. Despite these potential improvements, however, little gain in the pace of construction is expected in the next few years.

Final regulatory matters. The final review and licensing procedures are not critical factors in the time scale, since this review proceeds with construction. The process is facilitated by rigorous requirements for quality control during construction, which were introduced by the AEC in 1970-1971, and strengthened in 1972-1973. These requirements are expected to cause difficulties for the next two or three years until experience is gained, but not thereafter. Of course, public intervention is a possible cause of delay at the final stage of plant construction as well as at the start.

Alternative Technology. Experience with the high-temperature gas-cooled reactor is limited to one small (40 MWe) operating plant, and one moderate-sized (330 MWe) plant which will begin operation in the fall of 1974. Due to limited experience, construction time is longer than that for a light-water reactor of similar size, and there is little hope of reduced construction time in the next five to eight years. The HTGR uses a uranium-carbide fuel array which is very different in composition and construction from LWR fuel elements. Special production and fabrication facilities are needed; plants that produce LWR fuel cannot be adapted to produce HTGR fuel. A fuel-fabrication plant is under construction, with completion expected by 1978, but it cannot produce more than six reactor cores by 1982.

The breeder program is eight years from a demonstration plant, with an additional ten years before commercial operation. European programs, particularly the French effort, are five to eight years ahead of the U.S. The breeder program is the ultimate nuclear fission energy source, but in the U.S. its influence will not be felt until 1990 or later.

Uranium Resources. Another important factor in a rapid expansion of the nuclear industry is the availability of uranium enrichment capacity. Natural uranium contains over 99 percent nonfissionable U^{238}; only 0.71 percent is the fissionable isotope U^{235}. In order to reduce the size and improve the economics of LWRs, it is customary to enrich the U^{235} concentration. The gaseous diffusion process is used by the U.S. and other nuclear powers for enrichment, but gaseous diffusion plants cost billions of dollars. An alternative concept based upon gaseous centrifugal separation has been developed in Europe; plants are now under construction for commercial operation, and orders are being accepted for separative work. Presumably, the process is competitive with gaseous diffusion, but economic details and a time schedule are not available.

The present capacity of America's three diffusion plants is 17.8 million swu per year. (The capacity of enrichment facilities is denoted in "separative work units" or swu.) Projections of nuclear generation in 1980 range between 85 and 112 thousand MWe, which will consume between 11.3 and 14.2 million swu per year, nearly the present capacity. Short range programs can increase capacity to about 27 million swu per year. Such capacity will be required by the mid-nineteen-eighties.

Synthetic Fuels

Substitute natural gas (SNG), synthetic crude petroleum (syncrude), and methyl alcohol (methanol) can all be produced from coal and from oil shale. Studies of such processes have not been conducted on a large scale in this country until recently, because of the domestic abundance of cheap natural gas and petroleum. Intensive work on coal-based processes was carried out in Europe prior to 1945, however, and it resulted in plants which supplied Germany's wartime fuel needs. After World War II, development in Europe also stopped, because of the availability of cheap foreign crude oil.

As domestic supplies of natural gas and petroleum have lagged over the past few years, interest in synthesis of fuels from coal and shale has revived, resulting in many different processes developed by various organizations. Most of the "second generation" coal-based processes for SNG and syncrude, however, are still in pilot stages at best. Design and construction of large plants, using one of these second generation processes without the benefit of further pilot-plant experience, would involve serious risks. Operations might limp along at a fraction of design capacity, for example, and could thus incur very high costs.

Only for one or two of the new processes could large-scale plant designs be undertaken today with anything approaching confidence. The newer coal-based processes for syncrude are still in their developmental stages. The technologies of methanol synthesis from coal and syncrude synthesis from oil shale appear to be in fair shape, and oil shale studies have progressed to the point where both Union Oil and the Colony Oil Corporation have announced tentative plans for new plants in Colorado, with start-up possible by the end of the decade. Synthesis of SNG directly from oil shale is still in the early developmental stages.

Of course, plants for conversion of coal to syncrude could be designed using the old European technology, which is known to "work." However, these old coal liquefaction processes appear too outmoded to deserve consideration. The old art for making SNG from coal is not severely outmoded, and is being carefully reviewed in this country, even though newer processes under development promise significant improvement.

For most processes, the products produced will find ready market acceptance. The characteristics of SNG match those of natural gas. Syncrude from oil and from coals are acceptable feedstocks for refineries. Coal-derived syncrudes, however, tend to contain a higher proportion of "aromatics" (molecules containing benzene rings), which are valuable for gasoline use, but not as suitable for Diesel fuel. Methanol is a newcomer when viewed as a fuel, and its exact position in the market remains to be established. It could be simply and cheaply converted to methane, and with more difficulty, to gasoline. It also could be used as a boiler furnace fuel. Its use as a gasoline additive remains to be evaluated.

The various techniques for making fuels from coal and oil shale are difficult to compare for several reasons. The processes often produce very different mixes of products, for example. Some simultaneously make SNG, syncrude, and a coke-like material called char, in proportions varying from one process to the next, or in proportions which can be varied at will over a considerable range within a single process. Others make primarily SNG, or primarily syncrude.

All the synthesizing processes involve environmental problems. To attain low operating costs, the proposed plants will be enormous in size, and must be located where the necessary combination of coal or shale, water, and transportation facilities are available. Procurement of the necessary process water will be especially difficult in some places, though cooling needs may be moderated by appropriate engineering, and in some situations the mines may yield a significant fraction of the water required. Run-off waters from wastes (ash and spent shale) will carry dissolved salts, which also represent an environmental problem.

The associated mining facilities will be correspondingly large, especially where strip mining is practiced. Large quantities of coal ash or spent oil shale must be disposed of. The coals used will produce 5 to 20 percent of their weight in ash, while the spent shale will be 80 to 85 percent of the raw shale in weight and will occupy a volume up to 50 percent greater than the shale before oil is extracted from it.

Somewhat arbitrarily, hypothetical SNG plants are generally taken to have a production capacity of 250 million cubic feet per day, consuming perhaps 16,000 tons of bituminous coal daily. One hundred such plants would produce only a third of the country's current gas needs, but would consume all the coal now being mined in the United States. A 40,000 barrel per day syncrude plant, based on coal, is equivalent to a 250 million cubic feet per day SNG plant, in that the heating values of the products produced daily in the two plants are roughly equal. Such a syncrude plant would consume perhaps 10 percent less coal than the equivalent SNG unit. But one hundred such syncrude plants would produce only about a quarter of the country's current 15 million barrel daily consumption of crude oil.

Costs of Synthetic Fuel Plants. Tables 9 and 10 show the results of a survey of available information on the costs of fuel synthesizing processes. All costs come from the open literature; due to limitations of time and facilities, no independent cost estimates are presented in this study. The details of the preparation of these data are reported in the Appendix.

To facilitate comparison, all costs in the tables are for plants producing fuel with a heating value of 250×10^9 Btu per day—that is, 250 million cubic feet per day of SNG, 40,000 barrels per day of syncrude, or 12,500 tons per day of methanol. Plant cost estimates made and published in earlier years have been updated to allow for inflation. Plant investments have been put as nearly as possible on an equal basis by including the same allowances for contingencies, start-up, construction loans, etc. These plant investments do not include the mines (except in the case of the shale operations), nor do they include housing for personnel, but otherwise they are complete.

Capital investments are presented in Table 9. In view of the great uncertainties in predicting costs—especially the costs of new and unproven technology to be installed in plants of enormous size—these figures cannot be regarded as exact. In such situations, success in predicting capital costs to within 35 percent is unusual. Further uncertainties are introduced because the original cost estimates were made by different groups whose design philosophies inevitably differed. Finally, the estimates assume that the costs of future construction are adequately represented by estimates in 1973 prices. As noted later in this section, a rapid buildup of investment in these types of facilities could put severe strain on the construction industry, thereby driving up prices.

Table 9
CAPITAL COST OF SYNTHETIC FUEL PLANTS

Process	Capital Cost, in Millions of 1973 Dollars
SNG from coal, old technology	$400
SNG from coal, new technology	$300 to $350
SNG from oil shale [a]	$350
Syncrude from coal	$350
Syncrude from oil shale [a]	$450
Methanol from coal	$350

[a] The cost shown for the oil-shale processing plant includes an investment in mining and in waste-disposal facilities; the plant alone would cost perhaps $300 million.

Table 10
THE ANNUAL OPERATING COST OF VARIOUS SYNTHESIZING PLANTS
(millions of 1973 dollars)

	SNG from Coal Using Old Technology	SNG from Coal Using New Technology	Syncrude from Coal	Syncrude from Oil Shale [a]	Methanol from Coal
Capital, at 15 percent per year	59	44	51	37	51
Operating costs	22	16	22	22	44
Fuel costs	48	44	37	37	48
Total cost	129	104	110	96	143
Cost per million Btu of product	$1.56	$1.26	$1.33	$1.17	$1.73
Cost per barrel (oil equivalent)	$9.05	$7.30	$7.70	$6.80	$10.00

[a] The capital and fuel costs for the oil-shale plant reflect the costs of mining, crushing, and handling raw shale, and disposing of spent shale. The cost of coal is calculated using a price of 32 cents per million Btu.

52

While recognizing these limitations, it is worth noting that for all processes, predicted capital costs are very similar, falling into a range of $350 million to $400 million for the plant size considered.

Typical operating costs for SNG, syncrude, and methanol plants are presented in Table 10. These costs are subject to the same kinds of uncertainties as the capital costs of Table 9. All cost factors such as coal price, labor rates, maintenance expense, taxes and insurance, and so on are charged to the various processes at the same rates, as detailed in the Appendix. The 15 percent capital charge in these estimates is typical of published costs—it allows 10 percent for return on capital and 5 percent for depreciation. But a 15 percent rate is more representative of utilities financing than of private industry. Any increase, of course, will increase total costs of operation correspondingly.

Of particular note are the clustering of operating costs in the vicinity of $1.20 to $1.60 per million Btu, the sensitivity of total costs to the costs of coal or shale, and the fact that escalation of construction costs could drastically change these figures. Recognizing all this, it appears that shale oil costs, at around $1.17 per million Btu, are lower than those for the other processes considered.[1]

Synthetic fuel enterprises would be highly capital intensive. A 40,000 barrel per day syncrude plant, for example, would require an investment of around $350 million and, with a product price of $10 per barrel, would have an annual sales income of $130 million. The ratio of capital to annual sales is thus approximately three to one. A similar ratio of plant investment to probable sales income is expected for SNG and methanol. In the chemical industry, however, one dollar of plant investment usually generates about one dollar of annual sales.

In themselves, high ratios of capital to annual sales are not necessarily discouraging. Plants with such ratios are attractive when there is assurance of sales income in future years that is adequate to cover all costs and yield an attractive return. This is attainable only when technological risks which might result in curtailed production are minimal, and when unit sales prices do not fall.

Construction Times for Synthetic Fuel Plants. For coal-based methanol plants and oil-shale processing units, it appears that design and construction could be undertaken at an early date with fair confidence that smooth operation would be immediately attainable. SNG plants by old (Lurgi) technology might also be placed in this category, if it were

---

[1] Due to lack of reliable data, it has not been possible to include comparison of the various *in situ* shale-processing techniques now under consideration.

certain that the coal to be used would be processed successfully in the Lurgi reactors. (American coals often become sticky, resulting in serious handling difficulties.) Planning of SNG plants using the most advanced of the second generation technology could be undertaken in perhaps two years. Syncrude from coal will probably require three to five years of work before plant designs can be undertaken with much confidence. SNG production directly from shale (direct hydrogasification) lies farther in the future.

If the necessary design and pilot data are available, plant construction times are now set at three years. An additional two to three years is normally anticipated to obtain the many permits and straighten out the red tape involved. Demonstration plants for SNG and syncrude should certainly be built soon to explore the problems of synthesis. It would, however, seem undesirable to be stampeded into constructing a large number of such plants based exclusively upon old technology. The argument is often advanced that because of inflation, plants using old technology, built today, will have no higher costs than plants using improved technology, built a few years from now. The flaw appears to be that as a result of such thinking, the country could be saddled with outmoded plants of lower coal or shale efficiency.

In view of the process uncertainties, the scale of the operations involved, and the many difficulties in plant location, it seems unlikely that any significant production of synthetic fuels can be expected much before the mid-nineteen-eighties at best. Even then, a tremendous effort would be required to replace a large fraction of natural gas and petroleum with fuels derived from coal and shale.

The Capacity of the Construction Industry. The volume of construction of nuclear and fossil electric plants, refineries, and other energy facilities needed in the next decade is formidable. If large-scale coal gasification and liquefaction or oil shale processing plants are to be constructed as well, the pressure on particular segments of the construction industry will be tremendous. Prices of inputs to this industry are likely to be forced upward, raising the costs of all these facilities. Bottlenecks will develop (they already exist in some areas), causing construction delays and disruption of orderly planning. Moreover, the cost estimates for new facilities quoted in Tables 9 and 10 are based on the assumption that construction can be scaled up without difficulty. To the extent that this assumption is incorrect, costs are understated. Unfortunately, the capacity of the construction industry to expand is not well understood at present.

To give an impression of the magnitude of the construction task, Table 11 draws together some rough figures on the construction of certain electric-power and fuels-processing facilities. The figures for electric power plants and oil refineries are based on current experience and represent typical values. Figures for oil shale, coal gasification, and coal liquefaction are estimates by knowledgeable industry representatives which, because no large-scale units have been built, must be considered as *very* rough—perhaps off by 30 percent in either direction. Nonetheless, the magnitude of the problems of design and construction are plain.

There are relatively few firms involved in the construction of plants like those listed in Table 11. As these firms attempt to scale up to meet requirements for larger volume, problems are likely to develop in three areas: manpower, materials, and machinery.

Manpower constraints. As an example of the pressures already existing and increasing, Stone & Webster Engineering Corporation this year will manage 18 million manual man-hours on the construction of electric power plants. This number will have to triple to 60 million man-hours in 1979 for work already under contract or in negotiation. In its offices, Stone & Webster now employs about 1,800

Table 11

EFFORT REQUIRED TO CONSTRUCT VARIOUS ENERGY FACILITIES

Plant	Years to Design and Build	Millions of Man-Hours for Engineer/Builder [a]	
		Technical	Manual
1,000 MW electricity from nuclear fuel	8.5	1.5	11
600 MW electricity from fossil fuel	4.5	.5	3
200,000 bbl/day oil refinery	4	1.5	11
40,000 bbl/day syncrude from oil shale	4	1.0	10
250,000 cf/day SNG from coal	5	1.5	10
40,000 bbl/day syncrude from coal	5	1.5	10

[a] Does not include time requirements for the owner and vendors.

engineer-draftsmen. This staff must also triple by 1979. The same situation, more or less, holds for Stone & Webster's competitors. According to Stephen D. Bechtel, chairman of the Bechtel companies: "Our in-house studies show that the growth in skilled craftsmen and the supply of engineers are not keeping up with demand." Bechtel, which is the largest designer and constructor of electric power plants in the U.S., must grow from 20,000 salaried employees today to 50,000 by 1980.

Insofar as engineer-designer manpower is concerned, it does not appear that transfer from aerospace or other industries to plant design is possible for more than a very small fraction of the specialists required. Even transfer between power plant design and petroleum plant design is difficult. Experience appears to be crucial to productivity in process design and development.

One solution to the technical manpower problem is "standardization." The vast majority of plants being designed today are one of a kind, but standardization may well increase in coming years. Yet it is naive to think that the problem will be solved in this way. This is particularly true for unproven technology, where processes and safety requirements change continuously, and where entrepreneurial risks are high.

The most critical trades in plant construction are those that work with steel—pipe fitters, boilermakers, ironworkers, and so on—and electrical installations. One of the most serious bottlenecks appears to be pipe fitters. On a 40,000 barrel per day syncrude plant the contractor will need about 500 pipe fitters at the peak. A coincidence of ten such plants under construction could require 5,000 pipe fitters. With existing union constraints, a problem exists. To get large numbers of such workers into Montana and Wyoming presents further difficulties.

Materials problems. The principal material used in chemical plants is steel: about 170,000 tons go into the tanks, piping, heat exchangers, and so forth in a typical 200,000 barrel per day oil refinery. (We have no firm data, but believe that copper—used in all the wiring and electrical equipment—is the next most critical basic material.)

Even today, procurement of steel is a bottleneck in plant projects. For example, the lead time for orders for steel reinforcing bars is greater than one year. Steel pipe also is in very short supply. Certain sizes are not available at all, and the ordering lead time for what *is* available is increasing rapidly. Some firms are now designing plants that use available pipe sizes rather than optimizing their design.

Vendor-fabricated items. In power-plant and oil-refinery design and construction, many major items are designed and fabricated by manufacturers. Examples are valves, compressors, nuclear steam supply systems, boilers, fractionating towers, and so forth. Under current pressures, many vendors are already working at capacity. To increase capacity, they must expand both their fabricating capacity (shop facilities and labor) and their engineering capacity.

Consider, for example, fractionating towers, which might be sixty to eighty feet tall and twelve to fifteen feet in diameter, and which have a complicated internal structure. Coal liquefaction plants having a total capacity of one million barrels per day might require twenty or thirty such towers. At the present time only about a half-dozen vendors make them, and each tower might require six to seven months of shop time. A requirement to produce thirty towers in a short period of time would necessitate large capital expenditures in increased shop capacity.

No matter what evolves as government policy, the subsector of the construction industry dealing with plant design, procurement, and construction will be operating under severe stress during the next decade. We have focused here on power plant and chemical plant construction. Similar problems may arise in coal mining, coal transport, and offshore oil drilling. Pressure for significant increases in output will cause critical problems concerning input factors such as process engineers, design engineers, draftsmen, alloy steel, foundries, engineered equipment, pipe fitters, electricians, and managers. No analysis is available to show what this pressure will do to the costs of major energy-producing facilities, but there seems little doubt that the more intense the push for domestic fuels, the more likely it is that estimates made in 1973 will understate the actual costs.

6

THE UNITED STATES AND THE WORLD OIL MARKET

The price of imported oil is an important factor in the analysis of U.S. energy policy. The world price determines the resource cost of oil from abroad, and one of the goals of independence is to avoid a large economic drain for energy imports. Moreover, the world price is an important determinant of domestic price and, thus, of the incentives to domestic demand and supply.

The world oil price is not determined by the interplay of demand and cost but by a combination of economic factors and the political actions of producer governments. In such a circumstance, the forecaster's lot is not a happy one, and the choice of a policy to control imports requires careful study. We will begin with a brief analysis of the world oil price and some of the forces that influence its movement over time. We then turn to a discussion of the possible use of tariffs and quotas to buffer the U.S. economy from the vagaries of this market. Finally, we look at measures, such as stockpiling, that can help soften the impact of any future interruption of the flow of oil imports to the United States.

The Future of World Oil Prices

Oil is the largest single item in international trade, and the markets in which this commerce takes place are diverse and complicated. The principal producing areas are detailed in Table 12, which shows that the Western Hemisphere now produces about one-quarter of the world's crude oil. The Persian Gulf currently produces 38 percent, and this percentage is growing, since the gulf countries contain over 60 percent of the world's proved reserves of crude oil. The principal net importers of oil are the United States, Japan, and the countries of

Table 12

WORLD OIL PRODUCTION AND PROVED RESERVES IN 1973

	Production		Reserves	
	Million bbl/day	Percent	Billion barrels	Percent
Western hemisphere	16.1	28.9	76.1	13.4
United States	9.2	16.5	34.6	6.1
Venezuela	3.4	6.0	14.2	2.5
Canada	1.7	3.1	9.7	1.7
Others	1.8	3.3	17.6	3.1
Western Europe	0.4	0.7	15.9	2.8
Middle East	21.4	38.3	350.3	61.7
Saudi Arabia	7.7	13.8	140.8	24.8
Iran	5.9	10.5	60.2	10.6
Kuwait	3.1	5.6	72.7	12.8
Iraq	2.0	3.5	31.2	5.5
Others	2.7	4.8	45.4	8.0
Africa	5.8	10.5	67.6	11.9
Libya	2.2	3.9	25.6	4.5
Nigeria	2.0	3.6	19.9	3.5
Algeria	1.0	1.8	7.4	1.3
Others	0.6	1.1	14.7	2.6
Asia-Pacific	2.2	4.1	15.9	2.8
Indonesia	1.3	2.4	10.8	1.9
Others	0.9	1.7	5.1	0.9
Communist countries	9.8	17.5	42.0	7.4
USSR	8.4	15.1	34.6	6.1
China	1.0	1.8	7.4	1.3
Others	0.4	0.7	—	—
World total	55.7	100	567.8	100
OPEC members	30.8	55.3	416.3	73.3
AOPEC members	18.4	33.0	299.8	52.8

Source: *International Economic Report of the President,* January 1973.

Western Europe. These and other consuming nations face producer governments that are organized into two cartel-like organizations: the Organization of Petroleum Exporting Countries (OPEC), which contains all the major exporters, and the Organization of Arab Petroleum Exporting Countries (OAPEC), which contains the Arab subset of OPEC.[1] At present, the price of oil on the world market is determined by the monopolistic behavior of these exporters.

The pricing arrangements for world oil have evolved over the years into a very complex system. For the purposes of this discussion, however, they were drastically simplified in December 1973, when the Persian Gulf nations declared a tax of $7.00 per barrel on "own oil"—oil produced by international companies under various concession agreements. (This figure is for a particular type of crude oil. Prices of other crudes are adjusted for quality differentials.) The cost of production in the Persian Gulf is about ten to twenty cents per barrel, and the cost of transport to the U.S. and estimated oil-company profits total about $2.00 per barrel. Thus the price of a company's "own oil" from the Persian Gulf, delivered to the U.S., was a little over $9.00 per barrel as of mid-1974. Other exporting countries (which may have production costs higher than ten cents per barrel, but lower transport costs) adjust their taxes per barrel to follow the lead of Persian Gulf exporters.

In addition to the companies' "own oil" from concessions, there is the so-called "buy-back" oil which the producing nations own, and which they sell to the oil companies. Some producing nations are attempting to collect even higher prices for this part of their output, and as a result oil prices have been extremely uncertain during the first three quarters of 1974, particularly in the Persian Gulf. Oil companies know what tax they pay on "their own" oil, and governments inform them what they must pay for "buy-back" oil, but the proportions of the two are changing from month to month. At the start of the year, "own" oil was about 75 percent of production in the Persian Gulf; it will probably turn out to be no more than 50 percent and could drop to zero before the year is out. Since the settlement of this issue is likely to be retroactive, it is literally true that companies have been selling oil whose cost to them is unknown. Since March, companies have been transferring oil to their subsidiaries at a price of about $9.50 per barrel, which no doubt includes some allowance for contingencies.

---

[1] The members of OPEC are Algeria, Ecuador, Indonesia, Iran, Iraq, Kuwait, Libya, Nigeria, Qatar, Saudi Arabia, United Arab Emirates, and Venezuela. The members of OAPEC are Algeria, Bahrain, Egypt, Iraq, Kuwait, Libya, Qatar, Saudi Arabia, and United Arab Emirates.

Our assumption at other points in this study has been that the average payment will converge in the short run on a figure close to $7.00 per barrel (in 1973 dollars). That is, the Persian Gulf nations may be unable, as a group, to agree on any higher price level; their interests diverge on this issue, and within the cartel there has been opposition to increases. Under this assumption, the price of Persian Gulf oil delivered to the United States would remain near $9.00 per barrel. Obviously, such a prediction of Persian Gulf oil prices for the near future is no more than our attempt to foresee whatever order may emerge from the current confusion.

Looking farther into the future, the uncertainty inevitably increases. Naturally, the ability of producers to sustain the price at a level many times greater than cost depends on their ability to keep production from outstripping demand at that price, but at this point alternative visions of the future begin to diverge widely. Depending on one's opinion about the ability of the cartel (or key members of the cartel) to restrain production, one can make very different estimates of the future price of oil. If producing nations can perform as a classical cartel and effectively control production, they can sustain any price they desire (subject in the long run to the possible development of substitutes for crude oil). Or, if one country has a sufficiently large share of the market, and is willing to absorb all the required reductions in output (as some argue that Saudi Arabia is), then not even the cartel is needed; one producer alone can sustain the market price above the competitive level.

Analysis of such a situation is difficult, for the essence of a cartel is its unpredictability under strain. Nonetheless, even on very conservative assumptions it can be predicted that the level of strain could become very great between now and 1980. The price in the Persian Gulf has increased nearly eight-fold since 1970, from $1.25 per barrel to the current and uncertain $9.50. (It is important to look back three or four years in considering price increases because of lags in response.) Even if the $9.50 price falls toward $7.00 per barrel, there remains a five- to six-fold increase, to which producers may respond by increasing production. (Taking into account the much smaller rise in transport costs, the price increase at the oil's destinations is roughly four-fold.)

The average growth rate in world petroleum demand from 1962 to 1973 was 7.5 percent per year. If this pace of growth were to continue through the nineteen-seventies, world oil demand would increase from 55.7 million barrels per day in 1973 to 92.5 million barrels in 1980. But this neglects the effect of the recent four-fold increase in prices. On the conservative assumption that the price elasticity of demand is —0.15, this price increase would reduce 1980 demand by about 19 percent, to 74.9 million barrels per day. (If anything, such a prediction overestimates the demand likely in 1980, since studies of petroleum demand indicate long-run elasticities in the range of —0.2 to —0.4. Moreover, in the current situation, strong and often drastic demand-reduction measures are planned by governments throughout the developed and underdeveloped world—thus adding to the normal price responses of industries and households.)

A rough approximation of what could happen to supply is shown in Table 13. The potential U.S. supply is the same estimate shown in Table 1, with a deduction of two million barrels per day of natural gas liquids. Other non-OPEC countries are assumed to have a supply elasticity of 0.35, which is less than the implied coefficient for the U.S. despite this country's status as a special and unfavorable example

Table 13
ESTIMATE OF WORLD OIL PRODUCTIVE CAPACITY IN 1980

	Production in 1973, in Millions of Barrels Per Day	Estimated Productive Capacity in 1980, in Millions of Barrels Per Day
Non-OPEC	24.9	42.7
United States	9.2	8.7
North Sea	.0	6.4
Other non-OPEC	15.7	27.6
OPEC "expansionist" nations	12.2	17.9
Algeria	1.0	1.1
Indonesia	1.3	2.0
Iraq	2.0	4.0
Iran	5.9	8.0
Nigeria	2.0	2.8
OPEC "conservative" nations	18.6	24.9
Venezuela	3.4	3.4
Kuwait	3.1	3.3
Libya	2.2	2.2
Other Persian Gulf	2.2	2.8
Saudi Arabia	7.7	13.2
World total	55.7	85.5

of increasingly depletion-restricted output. For the North Sea, the estimate is that of an oil company working in the area, and includes the oil equivalent of natural gas production.

OPEC producers do not appear to be all alike in their plans and desires for increasing oil output. For purposes of this simple exercise, we divide OPEC into two rough subgroups: those countries which can be expected to increase output as rapidly as possible—denoted "expansionist" in Table 13—and those which will not increase production nearly so rapidly as they are able to—denoted as "conservative." Insofar as possible, the estimates of output in 1980 reflect each country's announced production plans. For the "expansionist" group, the implied price elasticity, for a five-fold price increase, is only 0.24. Outside Saudi Arabia, anticipated production expansion by the "conservative" group is almost nil.

Totaling these rough estimates, world productive capacity in 1980 is forecast to be 85.5 million barrels per day, which is 10.6 million barrels greater than the forecast demand—an excess of 16 percent. This excess is approximately 27 percent of potential OPEC production. Furthermore, the predicted supplies of 42.7 million barrels per day from non-OPEC countries and 17.9 million from "expansionist" OPEC countries leave a market of only 14.3 million barrels per day to be supplied by the "conservative" OPEC group—a quantity which is actually less than their 1973 sales of 18.6 million barrels per day. In fact, under these assumptions, the "conservative" group could just about meet demands for its production if all but Saudi Arabia held production at 1973 levels, and Saudi Arabia ceased production altogether.

To make this calculation, we have assumed that non-OPEC oil will be preferred to OPEC oil in most non-OPEC countries, so the demand for OPEC oil could be thought of as the difference between the total world demand and the non-OPEC supplies. If this and the other assumptions that lie behind the calculation are plausible, then the stability of the current price is questionable so long as Saudi Arabia must be depended upon to exercise all production restraint. In such a situation, Saudi Arabia may well see a somewhat lower price as being in its own interest. On the other hand, it is possible that the cartel could enforce production restraint on all or a significant number of its members. Certainly it is in the interest of all the producers to do so, but the history of commodity cartels does not give grounds for confidence that OPEC will be successful.

The primary conclusion to be reached from our thought-experiment is that the range of conceivable values for the world oil price over the next decade is very large. The price could rise above the current level (in 1973 dollars) through the efforts of the dominant exporters. On the other hand, the price could fall, due to the Persian Gulf countries' perceptions of their own long-run interest, or to a succession of price shadings and a failure of understanding, powerfully aided by attempts of buyers to obtain long-run contracts at lower prices. Rough calculations and study of recent data indicate that the price is more likely to fall than rise over the next few years, but the hard fact is that U.S. policy will have to be set in the face of inescapable uncertainty about this key parameter.

The U.S. can, of course, implement policies to buffer its economy from the world market. It is to these options that we now turn.

Tariffs and Quotas as Instruments of Oil Policy

For any tariff, there is a theoretically equivalent quota. In other words, if one knows the supply and demand responses to price changes, one can predict the level of imports under a certain tariff, and simply allow that quota to enter the country. Even the revenue-creating effect of a tariff can be duplicated by auctioning off the licenses available under a quota, thereby transferring to the U.S. Treasury the same funds which would be collected under a tariff. For oil, however, we lack the information about supply and demand which is necessary to assure the equivalency of particular quotas and tariffs. This lack, of course, is particularly acute with respect to world oil prices, since the overwhelming portion of that price consists of producer-country taxes which could be compressed without making oil unprofitable to sell.

Under these circumstances, tariffs and quotas thought to be the same will have divergent results, depending upon future movements in supply and demand. For instance, if world prices decline unexpectedly, a tariff will result in an unanticipated increase in the percentage of the domestic market supplied by foreign oil, whereas a quota will lead to an increase in the value of an import license but will leave the level of domestic production unaffected. Conversely, the failure of domestic supply to expand as expected will lead under a tariff to an increase in imports, but under a quota will cause an unanticipated price increase for domestic oil.

Under either a tariff or a quota, it is possible to make periodic adjustments as information becomes available. Indeed, except by pure luck, some adjustments will inevitably be required. But the nature of the adjustments will differ under a quota and a tariff, and the political and institutional constraints on adjustment-making will consequently be different.

Tariffs. The major advantage of a tariff over a quota is the maintenance of competitive pressure from the world market on domestic prices. When world supply is elastic, any attempt to increase domestic prices is limited by the potential loss of markets to foreign production, and foreign competition remains a spur for domestic efficiency.

A simple tariff, however, cannot limit the country's exposure to imports with assurance. As we have noted, if world prices drop, imports will be greater than expected and domestic production will decrease. This particular difficulty can be avoided through some sort of variable tariff, which could be set at the difference between some targeted domestic price and a base foreign price: for example, the average tax-paid cost of crude oil at the Persian Gulf, adjusted by a standard transportation cost. But this converts the tariff from a device that puts a price ceiling on domestic production to a device that sets a price floor. Doing so removes the competitive pressure of imports and thus neutralizes the major advantage of the tariff.

From a security point of view a variable tariff is also weak, for two additional reasons. First, it still does not protect the country from an undue penetration of imports if either the domestic supply or demand curve is mis-estimated. (This problem can be avoided by raising the tariff to achieve a targeted import level, but this makes it a disguised quota.) Secondly, adjustment as information becomes available may prove politically difficult: downward adjustment will cause protests by oil producers, and upward adjustments will cause protests by consumers. As a result, changes are likely to be delayed, and inertia may prevent any change at all. Political resistance might be avoided if the tariff were levied as a formula (like the one mentioned in the last paragraph) rather than a dollar figure, and recalculated regularly. A formula, however, could not take care of errors in the estimation of domestic production and demand unless it was expressed in terms of import volumes, in which case the tariff again becomes a quota.

A variable tariff which establishes a floor for domestic prices might have an advantage over a quota, to the extent that it gives domestic producers a clearer guide to the future course of domestic prices, and thus gives them greater certainty in evaluating their investment prospects. A quota that is apparently equivalent to a particular tariff in its protective effect may not result in the same level of investment, if firms are risk-adverse.

Quotas. The major advantage of a quota is its ability to limit directly the country's exposure to foreign oil. An oil import quota was in effect in the U.S. from 1959 until the early nineteen-seventies, but it was defective for a number of reasons. First, the quotas were imposed in a time when domestic production was heavily influenced by the "market demand prorationing" practiced by major oil-producing states—most notably Texas, through the regulatory actions of the Texas Railroad Commission. Under prorationing, these states were able to fix the total output by assigning a production quota to every well within their borders. Thus, in effect, control of U.S. oil prices passed to the Texas Commission, which attempted to manage output in order to maintain prices. This created substantial excess capacity and inefficiencies in domestic production. Secondly, the import licenses were distributed to refineries in a fashion that provided limited benefit to consumers and no revenue to the Treasury. Finally, the quotas were subject to a large number of exceptions unrelated to security objectives.

It is possible to have a pure quota which is not subject to these defects. Under such a quota, import rights would be auctioned off to buyers periodically, the revenues from the auction accruing to the Treasury. Anyone could buy licenses, and they could be resold. Since various middlemen could be involved in these transactions, it might be possible for the true identity of any buyer to be kept secret, in case he wished to evade an agreement not to bid. Both long-term and short-term quotas could be sold; the former would allow importers to make investment decisions with assured access to the U.S. market, and the latter would take care of the needs of the spot market.

One of the advantages of a secret competitive auction of import licenses is its potential for capturing some of the monopoly rents now being paid to the exporting countries. Bidders for import rights who must themselves purchase oil will bid only the difference between world and U.S. prices. Producer governments, on the other hand, could bid more, under threat of losing sales in the U.S. market, and transfer the licenses to producing companies. The true cost of oil to most exporting governments, particularly those in the Persian Gulf, is near nothing, and the cost to national oil companies is only ten cents to $1.00 per barrel. Sales in the United States could be enormously profitable, and thus the national companies could afford to bid an amount for import rights that is greater than the spread between world and domestic prices.

The described bidding procedure has more general advantages. It offers the prospect of capturing more rents for the Treasury than would a tariff (which of necessity would be set at the difference between the world and domestic prices). It also encourages competition among OPEC members, and any competition could put stress on the cartel and in the long run lead to an erosion of prices. Even if competition for the U.S. market does not emerge, the quota would produce the same revenue as a tariff.

One problem with a quota has been the difficulty of enforcing it when there is upward pressure on domestic prices. Conversely, there may be pressure to tighten the quota if domestic prices begin to fall. From 1971 to 1973, at least, the U.S. government found it easier to relax the quota than to meet the political pressures caused by a price increase. This may have been due to the historical use of the quota to protect producers, and to the presence in the early part of this period of excess capacity created by market demand prorationing, which made the necessity of a price increase look suspect. The recent crisis may make consumers more tolerant of price increases if they are necessary to limit imports. Nevertheless, this historical weakness should be noted.

Security Penalties. Even with a quota, it is possible to favor production from certain countries. Particularly secure sources of supply, such as Canada, could be exempted entirely, but doing so would allow either the Canadian government (through an export tax) or Canadian producers (through increased prices) to capture all of the difference between U.S. and world prices. There is no apparent need for the United States to be so generous. Even when U.S. and world prices are equal, Canadian oil can earn more in the United States than abroad because of the cost of transport.

Imports from sources deemed insecure could be charged a security fee. In the short run, when all supply sources are fixed, the fee could be avoided through exchanges among importers, substituting oil from "secure sources" for oil from "insecure sources." Such exchanges would not affect world output or the incomes of particular countries. But in the longer run, the burden of the fee would fall upon the producing country if competition for the U.S. market by national companies develops, or if there is insufficient oil from secure sources to satisfy the quota. In light of the large rents currently incorporated in world oil prices, an insecure country could still find it profitable to bid for access to the U.S. market, but to do so the country would be forced to share an even larger part of the rent with the U.S. Treasury.

If the fee were emulated by other consuming countries, its impact would be greater. In that case, even if there were no competition among producing countries for incremental sales, oil from countries deemed insecure would suffer a disadvantage if importing countries not imposing the fee were inadequate to absorb the desired level of production by insecure sources. As long as the supply of oil from countries considered secure was elastic, the burden of the fee would fall upon the producing country. The same effect could be achieved under a tariff system by setting different tariff levels for oil from secure and insecure sources.

Petroleum Stockpiles

Whatever policies are followed, the United States will be importing petroleum from potentially insecure sources for the remainder of the decade, and there is a good chance that such imports will continue for many years beyond that. Several measures could be taken to increase the flexibility of the country to deal with disruption of supply from the international market. They include emergency preparedness for curtailing consumption of petroleum products, measures to increase our capacity to convert from petroleum to other fuels during a disruption, provision for emergency increases in petroleum production from domestic reserves, and stockpiling of emergency supplies. All these precautionary actions deserve careful study. Their potential importance as elements of the U.S. relation to the world oil market can be seen in a very brief assessment of one of these measures—stockpiles.

The U.S. alone among major industrial nations has no provision for stockpiling crude oil beyond normal inventories. (It does maintain strategic stockpiles of many other natural resources, especially metals.) Our analysis suggests that stockpiling could be an important component of overall U.S. policy for ensuring a steady supply of energy. It would permit the U.S. to take advantage of a lower world price of oil without the costs and risks of relying on the timely arrival of imports.

The size of the stockpile depends on the magnitude of the expected interruptions and on the volume of imports permitted. For the sake of discussion, we will outline a "low" stockpile program that would permit a level of dependence of 15 to 20 percent on imports and complement other policies for achieving independence, and a "high" stockpile program that would permit unrestricted importation and would be the sole federal policy for independence. In both cases, we measure the costs of the programs at current prices and current levels of energy consumption. Future costs would rise in proportion to changes in these two numbers.

The Low Stockpile. Here we assume that some form of tariffs and import quotas, coupled with a stimulus to domestic production, serve to reduce U.S. imports to about 4 million barrels per day, and that half of this comes from insecure sources. We further assume that replacement of insecure sources may take approximately one year. Under these conditions, the stockpile should be about 730 million barrels. Estimated costs per barrel of above-ground storage facilities range between $3.00 and $5.00; costs for storage in underground cavities, such as salt domes, are about $1.00 per barrel. Some mix of these two types of facilities could be used in any large-scale stockpile program, and therefore the overall average construction cost for storage facilities should be approximately $3.00 per barrel. We further assume an annual operating cost of ten cents per barrel, and that the oil itself would cost $8.00 per barrel. Taking interest and depreciation of the facilities as 15 percent per year, and interest on the oil as ten percent per year, we find a total carrying cost of $1.35 per barrel per year. The annual cost of the low stockpile would therefore be $990 million.

European countries administer stockpiles by requiring oil companies to maintain them as a condition for the right to sell petroleum products. If this approach were adopted in the U.S., costs of oil companies would rise by about twenty-five cents per barrel of products sold, or about two-thirds of a cent per gallon. Alternatively, the storage requirement might be imposed on the importers of oil, in which case their costs, and prices, would rise by about 1.6 cents per gallon.

The low stockpile would require the construction of facilities costing $2.2 billion, phased over a few years. The construction methods required are conventional and should place little strain on the construction industry.

The High Stockpile. Here we assume imports of 4 million barrels per day from insecure sources, and assume further that three years would be required to develop alternative sources in the case of an interruption. The high stockpile would make the cost of interruptions like the recent embargo extremely high to the producing country, since it would forgo all revenues from the U.S. for three years before the interruption imposed any penalty on this country. The cost of this program is approximately six times the cost of the low program, or

$5.9 billion per year—perhaps somewhat more, because of the strain it would impose on the construction industry. Construction of the facilities and accumulation of oil would be phased over a longer period than for the low program. Costs and prices for all petroleum products in the U.S. would rise by $1.50 per barrel on the average, or 3.6 cents per gallon.

The cost of the high stockpile program compares favorably with the alternative of eliminating dependence on imports. Under the high stockpile program (with no restriction on imports), the price of oil would be $10.56 if the world price (delivered to the U.S.) were $9.00, whereas the U.S. price under import prohibition could be as high as $13.00. Consumers would save $2.44 per barrel, or about 6 cents per gallon. Moreover, U.S. natural resources would be conserved for later use when cheap foreign sources may be exhausted.

Many issues remain to be resolved about how such an emergency stockpile might be developed and managed, both in normal times and in an emergency. Nonetheless, the costs are sufficiently low, in relation to the likely costs of complete energy self-sufficiency, that this possibility deserves careful consideration.

7
POLICY CONCLUSIONS

Two goals underlie the emerging set of administration policies that have been termed Project Independence: (1) reducing the risk and the cost of disruption of oil imports and the attendant effects on foreign policy and the domestic economy, and (2) freeing the U.S. from the burden of high-priced oil imports, if the cost of imports is higher than the cost of increased domestic supplies.

Much discussion of Project Independence treats these goals as if they were one, but they are not necessarily even compatible. For example, if domestic supplies do not increase appreciably when the price rises, or if world oil prices decrease, then self-sufficiency could be bought only at the cost of an increase in prices *over* what might well be the long-term normal price of imports. In such a situation, the choice of policy measures depends on the relative importance attached to the two goals and on judgments about the critical uncertainties named in Chapter 1.

In many areas, very difficult choices must be made in the face of these uncertainties:

1. The way we normally solve domestic shortages is to let higher prices (perhaps supported by tariffs or quotas) suppress demand and bring forth increased domestic supplies. But because of the size of the energy industry and the magnitude of contemplated price increases, allowing these market forces to operate will involve significant transfers of income from consumers to the owners of energy-related assets.

2. Rapid development and utilization of coal will run into conflict with recent gains in the field of environmental protection.

3. The more quickly we try to achieve "independence," however it is defined, the greater will be the cost, in both economic resources and in environmental degradation.

In the face of such difficulties, the determination of firm policies requires analysis beyond that which we have been able to accomplish, and social judgments that ultimately must be made by responsible public officials. Nevertheless, the available information and our analysis of it reveal important features of the energy problem, and the policies that are likely to prove most effective and equitable in dealing with it. These measures fall into four areas: corrections to current federal regulatory policies, actions to let the market work effectively, actions to redress adverse effects on income distribution, and provision of security against disruption of oil imports, whatever the import volume may be.

Seek Revisions in Current Regulatory Policies

Many aspects of federal and state law and regulatory procedure serve to retard the growth of energy supplies and encourage the waste of energy. A continuing effort should be devoted to ferreting out these situations and, where possible, correcting them, even without the pressure of Project Independence. Naturally, most regulations have a good reason for existing, or had such at their inception, and change usually runs into conflict with the interests of particular groups or with some broader public concern. In several areas, revision will come with difficulty if at all, but they deserve to be noted nonetheless.

Field Regulation of Natural Gas Prices. Under current law and regulatory procedure, natural gas is seriously underpriced, leading to wasteful consumption and a reduction of supply incentive. Tables 4, 5, and 6 make clear the importance of correcting this imbalance. Field price regulation should be eliminated in phases over the next five years, along lines discussed in Chapter 3, as has been proposed in both the House and Senate in the last year. This change could make an important contribution to any program to approach energy independence.

Railroad Rate Regulation. One key to the effective use of western low-sulfur coal is the price of this coal delivered to markets in the East and Midwest. The outcome of rate cases under Interstate Commerce Commission jurisdiction can have a significant effect on this price. If, under the pressure of increased coal traffic, rates are allowed to rise too high (perhaps in order to subsidize other less profitable rail operations), coal use will be penalized and substitution of coal for oil and gas will be retarded.

Leasing of Federal Lands. By far the greatest increase in domestic energy supplies between now and the early nineteen-eighties will come from sources under federal jurisdiction, including offshore oil and gas, Alaskan oil and gas, and western coal. Given the long lead times for exploration and development, it is important that lease availability itself not continue to be a constraint, particularly in the case of offshore oil. The key here appears to be a rapid resolution of disputes over the adequacy of environmental safeguards.

Sulfur Pollution Standards. Beyond these first four issues lies a direct and unavoidable confrontation between the Clean Air Act (and associated implementation plans) and any scheme to substitute eastern coal for oil and gas. There is even doubt about the environmental consequences of burning much of the *low*-sulfur coal available in the West. Effective stack-gas desulfurization at a reasonable cost would cut the Gordian knot, but progress in this area is not a certainty. Our studies have not included analysis of the ways in which the Clean Air Act might be modified to allow increased coal use, yet preserve desired quality standards in our ambient air. It is clear, however, that any move away from current procedures will require much greater attention to ambient monitoring and to flexible systems of fuel control.

Licensing of Nuclear Power Plants. The present time cycle required for the design, certification, and construction of nuclear power plants affects the rate at which nuclear power can be substituted for the use of fossil fuels. As discussed in Chapter 5, changes in regulations to speed the approval process and decrease construction delays and costs are possible and needed.

Allow the Market to Work at Current International Prices

When the current confusion about Persian Gulf oil prices is clarified, the price in 1973 dollars will probably turn out to be between $7.00 and $8.00 per barrel. With the addition of transportation cost, the price as delivered to the U.S. East Coast will be over $9.00 per barrel. It is argued in Chapter 6 that it is unlikely—though certainly not impossible—that the price of world oil to the U.S. will be above $9.00

by the end of the decade. It seems more likely—though far from certain—that the prices in 1973 dollars will decline from the $9.00 level over the rest of the decade. Although it is hard to imagine the price dropping below $5.00 to $6.00 per barrel, there is no meaningful price floor set by the actual cost of oil, which is less than 5 percent of current export prices.

The price of Persian Gulf oil delivered to the U.S. was about $4.00 per barrel just one year ago, and therefore the production incentive offered by future prices over most of the expected range is very great. The question is how best to design a policy to take advantage of those forces when they work well, and what to do if even this incentive proves inadequate.

Price Maintenance. One way to strive for energy independence would be to use flexible tariffs or quotas to hold the oil price at a level that would clear domestic markets without (or nearly without) imports by the early nineteen-eighties. The analysis of Chapter 2 shows that it very likely will require a price above $11.00 per barrel to achieve this. This is an extremely high price, and policies that lead to it probably cannot be sustained. Given the purely technical limits on development of substitute fuels, even with large subsidies, it is unlikely that "independence" can be attained by the early nineteen-eighties if it is defined in terms of a zero-net import requirement.

However, even if the import price cannot be raised to a market-clearing level, the question remains: Should the price of imported oil be held at some level above the expected import price of $9.00—say, at $10.00 to $12.00 per barrel? Our conclusion is that it should not, for the following reasons:

The current import price, and the price of "new" oil,[1] which follows it, are high enough to provide adequate incentive for oil exploration. (Natural gas is a separate problem, as noted in Chapter 3.) A still higher price will have only a marginal effect on exploration and production over the next few years.

At the current price level for imports, and the associated price of residual fuel oil, there is also ample incentive for coal production. At this price, the most immediate barriers to expansion of coal are to be found in environmental issues, in problems of transportation and government leasing policy, and in limited demand.

---

[1] "New" oil includes all oil from wells completed since 1971, or from increases since 1971 in production from pre-existing wells. "Old" oil, which remains under price control, is the remainder of domestic production.

As indicated in Chapter 5, no one knows with any accuracy what price incentives it may take to spur development of synthetic fuels, or whether incentives are needed at all. Therefore, in the early stage of this new industry it is preferable to use selective policy instruments, such as specific plant process or product guarantees or subsidies, which are more precisely attuned to the problems of large-scale synthetic fuel processes.

The effect on demand of a $9.00 energy price is poorly understood, and even less is known about prices above that level. In the short run, such a price would yield little other than an income transfer from consumers to producers. The political difficulties that flow from these income transfers need no elaboration.

In short, the import price is high enough, and there is little to be gained from raising it higher. Should the world price rise still higher due to actions of exporting countries, this conclusion is only strengthened.

Protection against Down-Side Risk. Should the real price of imports (in constant 1973 dollars) begin to fall below the expected $9.00 level, another set of questions arises. Should a floor be put under import price by flexible tariffs or quotas? If so, at what level? Further, should that floor be established now, or can this possibility be left an open question? Our conclusions are the following:

By the same arguments as those used in the paragraphs on "price maintenance," there should be adequate incentives for oil, gas, and coal supply at prices as low as $7.00 per barrel, and therefore there is little need at this time to set a floor on import price, provided that proper incentives are offered to the infant synthetics industry. Should the import price drop into the neighborhood of $7.00 per barrel, this conclusion would have to be reconsidered. In addition, more data will become available on supply and demand response to current high price levels, and this judgment can be adjusted in the light of that new evidence as well.

As noted in Chapter 6, the choice between tariffs and quotas is a complex one in the new world-oil situation. There is a clear need for further in-depth study of these measures, and evaluation of their flexibility in the face of the inevitable uncertainty in the world oil market over the years to come.

Domestic Price Controls. Domestic "old" oil is currently being held at $5.25 per barrel, while "new" oil is selling at $10.00 or above. Under current definitions, "new" oil is about 40 percent of total domestic supply, and that percentage will grow with time. Such a two-tier pricing system leads to distortion and waste when applied to essentially the same product, especially when new investment in old oil leases is desirable. On the other hand, removal of price controls will result in a sizable increase in revenues and profits to certain segments of the petroleum industry. (Similar problems arise from price controls on gas.)

In this circumstance, we recommend that price regulation of "old" oil be relaxed (perhaps in phases), but that this be done only if provision is made to capture a portion of the resulting income transfers, by means to be discussed momentarily.

Feedback on World Price Levels. It is widely assumed that reduced American imports will lessen the strain on world supply, tending to lower world prices, while continued increase in our imports will tend to raise prices. This would be true if world oil prices were determined by the strain on world resources and were equal to marginal production costs. But since prices are many times as high, lower American imports will not necessarily reduce world prices. The essential determinant of prices is the internal cohesion and discipline of the cartel. If the cartel is able to hold capacity at a level not much above consumption, there need be no downward pressure on price.

Incentives to the Synthetics Industry. Thus far, we have argued that no commitment should be made to hold the price of energy high enough to make room for SNG, syncrude, methanol, or shale oil. But if this approach is taken, some specific programs will be needed for these sources, since their development to pilot and commercial productions stages should not be delayed.

The best way to accomplish this is to identify a special class of new energy sources, and provide specific price guarantees for the output of the first group of commercial-scale plants. In a few years, a great deal more will be known about the synthesizing processes, and better informed judgments can be made about their costs and the general price level it takes to give incentive for their expansion. To handle the first round of plants, one of two schemes is recommended:

1. Potential synthetics suppliers bid for contracts to supply SNG or syncrude to the federal government at some future date. Contracts would be in lots of, say, 50,000 barrels per day, and would be at a fixed price per barrel (plus adjustment for inflation) for an adequate period of years (ten to twenty). Through this

bidding process, it can be determined whether and how much subsidy is required.

2. The federal government offers to purchase oil from new commercial plants at a price agreed upon by negotiation.

Many details of such schemes remain to be analyzed. But by moving now to offer a price floor for the critical first generation of plants—up to perhaps 500,000 or a million barrels per day equivalent—the federal government can insure that development takes place at the maximum rate without a premature commitment to subsidize or otherwise provide protection for a massive industry as yet unborn.

Measures to Increase Energy Productivity. There undoubtedly are many places in the economy where energy is not utilized efficiently, even at increased prices, due to some failure in the working of the market mechanism. The savings to be gained from correcting these failures by selective public measures should be considerable, and a continuing effort should be made to isolate these cases and, where possible, to treat them. Obvious places to look include increasing consumer information through labeling programs and measures to facilitate the substitution of capital (for example, insulation) for energy in space conditioning.

Other possibilities include various forms of controls on the energy efficiency of particular devices or on the availability of the devices themselves. Our study has not included a detailed analysis of this topic, but a cursory investigation indicates that much remains to be done by way of careful evaluation of specific proposals. As noted at the outset, it is not yet understood how energy use will respond to the recent dramatic rise in prices. It is quite possible that specific controls could be instituted, at some costs of administration and disruption of target sectors, which accomplish little more than was going to take place anyway. Even where this is not the case, ill-considered control schemes could involve costs beyond their value in terms of energy savings.

Finally, energy "conservation" could be fostered by increasing the prices of specific refined products, of natural gas, and of electric power—by means, for example, of selective excise taxes. Traditionally, the federal government has avoided the use of substantial excise taxes on refined products, a practice that is common in Europe. Once again, the implications of such a change on specific sectors of the economy is not well understood at present, particularly considering our ignorance of the effects of price changes that have already taken place, and much evaluation needs to precede action in this area.

Correct Adverse Effects on Income Distribution

When the price of world oil is driven up very rapidly by action of foreign producers, it is inevitable that there will be an accompanying rise in the revenues and profits accruing to persons owning energy-producing assets of one kind or another. Several of the policies recommended above—such as natural gas deregulation and relaxation of domestic price controls on "old" oil—will contribute to this income transfer. The amounts of money involved are large enough to raise real problems of social equity. For example, a $5.00 price increase on 10 million barrels per day of domestic oil and gas amounts to an income transfer of $18 billion per year. Current political controversies over crude oil prices are, in large part, a reflection of this underlying equity issue.

Several solutions to this problem have been proposed, including a variable excise tax on domestic crude oil and various forms of excess-profits taxes. The best answer, however, is apparently to bring oil companies back into the U.S. corporate income tax system. At present, with foreign taxes (which run in the neighborhood of $7.00 per barrel) being credited against U.S. taxes on overseas operations, and with the 22 percent depletion allowance, the corporate tax liability of many oil companies is vanishingly small. We have not conducted an analysis of the details of alternative changes to the special provisions of the tax code for various parts of the energy industry, but the simplest and most straightforward solution seems to be to treat energy corporations more nearly like we treat the rest of American industry, and thereby to capture some of the rising income flows in the normal tax system.

With such a change, two questions would remain: Is the income transfer, after taxes, sufficiently reduced to prove socially acceptable? And are corporations left with sufficient retained earnings to finance needed investments? We suspect that the answer is yes, but both issues need further study.

Provide Security against Import Disruption

Implicit in the preceding conclusions is the judgment that the cost of security is too high if it is sought solely by eliminating imports. In effect, the two justifications for independence from foreign sources—avoiding oil blackmail and cutting the resource cost of our energy—are contradictory if the target date is the early nineteen-eighties. The economic cost of additional domestic supplies, on this time scale, is above the expected economic cost of imports. As a result, the U.S. is likely to be a net importer of energy right through the nineteen-eighties, though in drastically reduced quantities, as a result of the policies discussed above. The question is how to reduce the risk of these residual imports.

Far too little attention has been given to the possibilities and costs of providing a stockpile of crude oil beyond normal inventories. While we have not examined the possible response of OPEC to this program, or the best means of insuring that inventories are built up to the desired level, we believe that this option must be given serious consideration as another way of reducing the risks and costs of interruptions in our energy supplies.

APPENDIX

The Origin of Estimates of Synthetic Fuel Costs

Table A-1 presents data taken directly from the open literature on the various synthetic fuel processes. Only the total capital investment figures have been altered; we give them in 1973 dollars. Coal costs are standardized to thirty-two cents per million Btu for eastern bituminous coal ($8.00 per ton at 25 million Btu per ton), and sixteen cents per million Btu for western coal ($3.00 per ton at 18.75 million Btu per ton). Plant costs are for a completely self-contained unit, and include everything except cost of the mine (unless otherwise noted) and housing facilities for workers.

Four different estimates for Lurgi-type coal gasification plants appear in the table. They show significant variations in capital costs and efficiency, even though the plants have the same capacity, and are based on the same technology. Details on the estimating procedures used are scanty, so reconciliation of the differences is not possible. It is probable, though, that Fluor's estimate, being the most recent, is the most reliable. (Here it should be noted that the tar by-product is apparently not credited to the operating cost or to the coal consumed.)

The second group of processes presented in Table A-1 are those coal liquefaction processes for which cost estimates are available: H-Coal, PAMCO, COED, and CONSOL. Note that the capacities of these plants differ widely, as do their mixes of products. Costs for only one methanol plant were found in the literature, as presented in Table A-1.

Published data on projected costs of the oil-shale processes are likewise scarce. The costs of syncrude obtained from shale liquefaction obviously depends on the assay of shale used, and are five to seven percent lower for material containing 35 gallons per ton than for 25-gallon-per-ton material. Processing costs depend significantly on the cost of disposing of spent shale, apparently assumed in these estimates to be nominal.

The data in Table A-1 are presented in condensed form in Table A-2, after the following recalculations to simplify process comparisons:

1. Plant sizes differ by a factor of three in Table A-1. Investments in these plants have been scaled to a capacity of 250×10^9 Btu per day in Table A-2 by assuming investments to vary with the 0.9 power of capacity. This exponent is conservative and assumes that, for large plants, capacity increases are made by multiplication of units. Before such scaling, all plants costs are corrected to 1973 dollars using the *Chemical Engineering Plant Cost Index*.

2. Total capital investment has been standardized to include the same facilities, such as power generation and pollution control capabilities. Capital costs exclude the mine except where indicated. These investment figures also include 15 percent for contingencies.

3. Operating costs are similarly standardized. Thus coal, labor, taxes, and insurance are charged to all processes at the same rates, as detailed in the footnotes to Table A-2. The plants are assumed to operate 330 days per year.

Table A-1

PLANT COSTS FOR VARIOUS SYNTHETIC FUELS, AT 1973 PRICES

	Output Fuel (billion Btu/day)	Daily Plant Output	Input Fuel (tons/day)	Thermal Efficiency (percent)	Total Capital Investment ($ millions)	Annual Operating Costs ($ millions)	Capital Charge, 15% TCI/Year (¢ per million Btu of product)	Operating Cost (¢ per million Btu of product)	Coal Cost (¢ per million Btu of product)	Product Cost (¢ per million Btu)
Coal gasification										
FPC Lurgi—bituminous	237.2	250 million c.f. pipeline gas	16,000	59.2	347	23.1	66.5	29.5	54.1	150.1
FPC New—bituminous	240.0	250 million c.f. pipeline gas	15,900	60.3	296	16.2	56.1	20.5	53.1	129.7
FPC Lurgi—western	238.8	250 million c.f. pipeline gas	20,800	61.2	313	24.6	59.6	31.2	26.1	116.9
FPC New—western	240.0	250 million c.f. pipeline gas	21,200	60.3	261	16.1	49.4	20.3	26.5	96.2
NPC Lurgi—bituminous	243.0	270 million c.f. pipeline gas	14,300	67.9	285	20.8	53.3	25.9	47.1	126.3
NPC Lurgi—western	243.0	270 million c.f. pipeline gas	19,200	67.6	241	18.6	45.1	23.2	23.6	91.9
NAE Lurgi	235.0	240 million c.f. pipeline gas			318		61.5			
Fluor Lurgi—western	252.1	257 million c.f. pipeline gas	23,900	56.2	427		77.0	24.5	28.5	130.0

Table A-1 (continued)

	Output Fuel (billion Btu/day)	Daily Plant Output	Input Fuel (tons/day)	Thermal Efficiency (percent)	Total Capital Investment ($ millions)	Annual Operating Costs ($ millions)	Capital Charge, 15% TCI/Year (¢ per million Btu of product)	Operating Cost (¢ per million Btu of product)	Coal Cost (¢ per million Btu of product)	Product Cost (¢ per million Btu)
Coal liquefaction										
NPC H-Coal—bituminous	240.0	30,000 bbl. syncrude 60 billion Btu fuel gas	13,000	74	260	26.8 [a]	49.2	33.9	43.2	126.3
NPC PAMCO—bituminous	180.0	30,000 bbl. "de-ashed prod."	9,600	75	187	13.4 [a]	47.2	22.6	42.7	112.5
Amoco COED—bituminous	405.0	250 million c.f. pipeline gas 29,175 bbl. syncrude	28,400	57	500	53.8 [a]	56.0	40.0	56.0	152.0
Foster-Wheeler CONSOL A-Bit	358.8	284.3 billion Btu liquid 74.5 billion Btu gas	20,200	71	309	36.0 [a]	39.1	30.4	45.1	114.6
Foster-Wheeler CONSOL B-Bit	421.0	282.1 billion Btu liquid 138.9 billion Btu gas	25,100	67	405	43.3 [a]	43.7	31.2	47.8	122.7
Low-Btu-gas from bit. coal										
NAE—150–400 Btu/c.f.	235.0		11,800–13,400	70–80	165–189		31.9–36.6		40.0–45.7	110–125
FPC—150–400 Btu/c.f.	235.0		11,800–13,400	70–80	189–191		36.6–36.9			

Methanol from coal

ORNL—bituminous	391	20,000 tons methanol	23,300	67	416	58.4 [a]	48.4	45.3	47.8	141.5

Oil shale

FPC gasification—25 gal./ton	250	271 million c.f. pipeline gas	94,500	74	Process: 292 / Mine: 156 / Total: 448	20.6 / 22.8 / 43.4				133
NPC liquefaction—25 gal./ton	600	100,000 bbl. syncrude	174,800	96	Process: 542 / Mine: 210 / Total: 752; Process: 542 / Mine: 250 / Total: 792	44 / 46 / 90; 44 / 50 / 94				102–108
NPC liquefaction—35 gal./ton	600	100,000 bbl. syncrude	124,800	96	Process: 455 / Mine: 150 / Total: 605; Process: 455 / Mine: 175 / Total: 630	37 / 33 / 70; 37 / 35 / 72				82–85

[a] Excluding coal.

Source: All data are taken from the open literature. The sources are the following: "Final Report—The Supply-Technical Advisory Task Force—Synthetic Gas-Coal," National Gas Survey, Federal Power Commission, April 1973; "U.S. Energy Outlook—Coal Availability," Report of the Fuel Task Group on Coal Availability, National Petroleum Council, 1973; "Evaluation of Coal Gasification Technology, Part I—Pipeline-Quality Gas," National Academy of Engineering, 1972; "Evaluation of Coal Gasification Technology, Part II—Low- and Intermediate-Btu Fuel Gases," National Academy of Engineering, 1973; Wen, C. Y., "Optimization of Coal Gasification Processes," R & D Report No. 66, Office of Coal Research, 1972; Siegel, H. M. and T. Kalina, "Technology and Cost of Coal Gasification," *Mechanical Engineering*, May 1973, pp. 23-28; Moe, J. M., "SNG from Coal via the Lurgi Gasification Process," Symposium on "Clean Fuels Firm Loaf," Institute of Gas Technology, Chicago, Ill., Sept. 10-14, 1973; Shearer, H. A., "The COED Process Plus Char Gasification," *Chemical Engineering Progress*, vol. 69, no. 3 (1973), p. 43; Foster-Wheeler Corp., "Engineering Evaluation and Review of CONSOL Synthetic Fuel Process," R & D Report No. 70, Office of Coal Research, February 1972; Michel, J. W., "Hydrogen and Synthetic Fuels for the Future," 166th National Meeting, American Chemical Society, Division of Fuel Chemistry Preprints, vol. 18, no. 3 (August 1973); "U.S. Energy Outlook—Oil Shale Availability," National Petroleum Council, 1973.

Table A-2

PLANT COSTS FOR 250×10^9 Btu PER DAY OF VARIOUS SYNTHETIC FUELS [a]

	Input Fuel (tons/day)	Thermal Efficiency (percent)	Total Capital [b] ($ millions)	Annual Operating Costs ($ millions)	Capital, at 15 percent per year	Costs, 330 Days/Year Operation (cents per million Btu of product)		
						Operating costs	Coal or oil shale cost	Total cost
Coal gasification								
Lurgi-bituminous [c]	14,700–17,900	56–68	334–390	21.4–22.2	60.7–70.9	25.9–26.9	47.1–54.1	135–150
Lurgi-western [d]	19,600–23,800	56–68	290–390	19.5–23.0	52.7–70.9	23.6–28.5	23.6–28.5	100–127
Coal liquefaction	13,200–17,500	60–75	233–373	22–35	43.4–67.8	26.7–35.0	42.7–56.0	112–166
Low-Btu gas								
Bituminous [c]	12,500–14,300	70–80	195–208		35.5–37.8		40–45.7	110–125
Western [d]	16,700–19,000	70–80	195–208		35.5–37.8		20–22.7	90–105
Methanol-bituminous [c]	14,900	60–67	279–364	44	50.7–66.2	54	53.4	158–174
Oil shale								
Gasification [e]	94,500	74	415		53.2	25.0	56.1	134.3
Liquefaction [f]	72,900	96	342–360		45.0	27.0	42.8–48.2	114.8–120.8
Liquefaction [g]	52,000	96	276–327		37.6	23.4	46.8–52.6	107.8–113.6

a In this table, the figures shown in Table A-1 are made comparable by recalculations described in the text of the Appendix.

b Includes onsites, offsites, auxiliaries, 5 percent start-ups, 15 percent interest during construction, and 7.5 percent working capital. The plant costs are normalized to 250×10^9 Btu per day of product by assuming these costs to vary with capacity to the 0.9 power.

c Thirty-two cents per million Btu, 25 million Btu per ton, $8.00 per ton.

d Sixteen cents per million Btu, 18.75 million Btu per ton, $3.00 per ton.

e Labor at $5.50 per hour, 2.5 percent for taxes and insurance on plant investment, 4.5 percent for maintenance.

f Twenty-five gallons of oil per ton of shale.

g Thirty-five gallons of oil per ton of shale.

Cover and book design: Pat Taylor